ENHANCEMENT EXERCISES

for

BIOLOGY

Byron J. Adams • **John L. Crawley**
Brigham Young University

MORTON
PUBLISHING

925 W. Kenyon Avenue, Unit 12
Englewood, CO 80110
www.morton-pub.com

Book Team

President	David M. Ferguson
Senior Acquisitions Editor	Marta R. Martins
Project Editor	Rayna S. Bailey
Associate Project Editor	Sarah D. Thomas
Assistant Editor	Trina Lambert
Production Manager, Interior Design & Composition	Will Kelley
Production Assistants	Joanne Saliger, Sarah Bailey

Printed in the United States of America

10 9 8 7 6 5 4 3 2

ISBN-10: 1-61731-533-8
ISBN-13: 978-1-61731-533-6

Library of Congress Control Number: 2016944548

Why We Wrote This Book

We are convinced that active learning modules provide the best opportunity for students to learn and experience biology. We are also convinced that inquiry-based exercises are effective because they require students to actively engage in processes of scientific discovery.

Our primary goal with this book was to create exercises that:

1. Challenge students at an appropriate level of intellectual independence, ranging from simple, guided inquiry to more thoughtful, open-ended, research-based activities.

2. Emphasize an appropriate level of content mastery, from non-majors to majors.

3. Reinforce course learning objectives delivered through lecture, readings, and other course activities.

Secondly, we wanted the exercises in this book to be designed such that that they can be readily incorporated into any type of learning environment without the need to make significant changes to the existing course structure—whether it be a traditional lab and lecture course, a lecture course without a formal lab, a flipped lecture format course (in-class activities, outside of class reading and preparation), or an online-only course (see next page for examples).

Lastly, we wanted this book to provide instructors more opportunities to get students engaged in exploring the natural world in ways that nurture a sense of wonder. As students are increasingly exposed to fewer opportunities to observe the natural world, we hope that our book provides students with more opportunities to experience the same awe that drew us to biology in the first place.

How to Use This Book

This book can augment any college-level biology course, from a non-majors general education course to a very rigorous course for life science majors. As such, it likely contains topics you don't normally cover in your class. The exercises have a significant range of difficulty and detail, and you may find that of the topics you do cover, some activities may go into greater detail than you require. Each exercise starts simple and moves to more complex and detailed concepts and content. The purpose for providing extensive range of breadth and depth is to allow you, as the instructor, the ability to assign all or a portion of an individual exercise or even skip certain exercises and activities if they are not applicable to your specific course.

This book has been designed so that the student can complete the assignments without any need for specialized lab equipment. The exercises can be completed by visiting local outdoor environments, or by using common items easily obtained at home or the grocery store. You do not have to change your current course. Simply assign the exercises in a way that best fits your curriculum. These exercises make great take-home assignments for courses without a laboratory component. They can be used in courses with assigned labs that have limited resources for equipment and supplies. They also provide the flexibility of allowing the student the option of doing them on his or her own time, eliminating the problems that arise as a result of a student missing an assigned lab time.

Each exercise begins with terms and definitions and moves to progressively more challenging activities that require higher levels of thinking. This makes it easy for you to pick and choose the activities, and components of activities, that your students need most.

We suggest instructors not get too focused on grading every single student response, but rather how well the student or individual has moved through or engaged in the process. The open-ended nature of many of the exercises places much of the responsibility for learning and understanding back on the student. Thus, the exercises are designed so that a quick glance at the returned assignment is all you need to see to know if the student is on track or needs extra help.

Biology is a visual, hands-on discipline. We believe that the optimal environment for learning biology should be equally visual and hands-on. *Enhancement Exercises for Biology* will provide students with the opportunity to develop the ability to critically think and work through the process of finding solutions to exercises and questions that will greatly enhance their classroom experience.

Lecture Course With a Lab

○ In addition to the pre-lecture assignment from their textbook, students can read the *Enhancement Exercises* background information and complete the first activity, defining important terms associated with the topic. This can be done prior to class, prior to lab, or during the first part of the lab.

Lecture Course Without a Lab

○ Incorporate one or more of the active learning modules into the lecture, as an added component to the existing lab activity, or as a stand-alone lab activity. Alternatively, assign students to complete the module(s) outside of class or lab, preferably prior to lecture.

Flipped Classroom

Online or Hybrid Course

○ After lecture or the end of their lab, students can complete the Synthesis activity.

Sample Course Syllabus:

Course: Biology—General Education course for non-majors with a lab
Topic: Photosynthesis

Before class assignments:
- Complete textbook reading assignment.
- *Enhancement Exercises*, Exercise 11: Read objectives and background information.
- Complete *Enhancement Exercises* Activity 1.

During lab:
- Students explore the relationship between light, photosynthesis, carbon fixation, and sugar production (Activity 2).

During lecture(s):
- Standard lecture on photosynthesis.
- Students perform the active learning activity for understanding photosynthesis.

After Lecture:
- Students complete Synthesis (Activity 5).

Lecture Course With a Lab

Lecture Course Without a Lab

Flipped Classroom

Online or Hybrid Course

○ In addition to the pre-lecture assignment from their textbook, students can read the background information associated with the lecture topic in *Enhancement Exercises* and complete the first activity, defining important terms associated with the topic.

○ Incorporate one or more of the *Enhancement Exercises* activities into the existing lecture. Alternatively, students can complete these outside of class, preferably prior to lecture.

○ After lecture, students can complete the *Enhancement Exercises* Synthesis Activity.

Sample Course Syllabus:

Course: Biology for Life Science Majors (no lab offered)
Topic: Control of Gene Expression

Before class assignments:
- Complete textbook reading assignment.
- *Enhancement Exercises*, Exercise 19: Read objectives and background information.
- Complete *Enhancement Exercises* Activities 1, 2, 3, 4, 5; turn them in at the beginning of class. This is a formative assessment where students are scored on completing the assignment and giving their best effort. Working in groups to complete the assignment would work as well.

During lecture(s):
- Standard lecture on gene expression.
- When discussing components of the lac operon, have students return to Activity 2.
- When summarizing the mechanics of lac regulation, have students return to Activity 3.
- When discussing positive and negative regulation, have students return to Activity 4.

After class:
- Students turn in Activities 5 and 6.
- Synthesis (Activity 6) can be scored and serve as a summative assessment for this topic.

○ In addition to the pre-lecture reading assignment from their textbook, students can read the background information in *Enhancement Exercises* and complete the first activity, defining important terms associated with the topic, prior to class.

○ Spend class time helping students work through each of the active learning modules. This can be done alone or in groups.

○ At the end of class, or as an after-class assignment, have students complete the Synthesis Activity. Students can complete this individually, or in groups.

Sample Course Syllabus:

Course: Flipped Classroom—Biology for Life Sciences Majors (no lab offered)

Topic: Control of Gene Expression

Before class assignments:
- Complete textbook reading assignment.
- Watch online lecture(s).
- *Enhancement Exercises*, Exercise 19: Read objectives and background information.

- Complete *Enhancement Exercises* Activities 1 and 2, turn them in at the beginning of class. This is a formative assessment where students are scored on completing the assignment and giving their best effort. Working in groups to complete the assignment would work as well.

During class period(s):
- Guide students, alone or in groups, through Activities 2-4.

After last class on this topic:
- Students turn in Activity 6 (Synthesis), which can be scored and serve as a summative assessment for this topic.

○ Augment your current curriculum, or use *Enhancement Exercises* as a stand-alone source for the active, inquiry-based learning opportunities that your students would otherwise have access to in a standard lecture or lab course environment.

Sample Course Syllabus:

Course: Biology—General Education (online)

Topic: Photosynthesis

Assignments:
- Complete textbook reading assignment.
- Complete homework assignments associated with the current course.

Augmented lab experiments:
- Students explore the relationship between light, photosynthesis, carbon fixation, and sugar production by performing simple experiments at home (Exercise 11, Activity 2).

Acknowledgments

Many individuals have assisted in the preparation of *Enhancement Exercises for Biology* and share our enthusiasm about its value for students of biology. We are especially appreciative of Kaitlyn Anne Morgan and Marci Shaver-Adams for their assistance in researching topics, identifying potential exercises, providing valuable input throughout the development process, and help with the instructor's manual. We express gratitude to our colleagues and students in the Department of Biology at Brigham Young University whose thoughtful insights and ideas inspired our approach and many of the activities developed for this book. We are especially grateful to Jamie Jensen, Jerry Johnson, Keoni Kauwe, Stephanie Burdette, Riley Nelson, Rick Gill, Bill Bradshaw, Laura Bridgewater, and Mike Whiting, as well as students Tim Davie, JT Bliss, Scott Peat, Bishwo Adhikari, Dana Blackburn, John Chaston and Adler Dillman. We also are very appreciative of Stephanie Burdett from Brigham Young University, Fleur Ferro from the Community College of Denver, and Jennifer Nauen from the University of Delaware, for detailed reviews and feedback concerning this project.

We are indebted to Douglas Morton, David Ferguson, Marta Martins, Sarah Thomas, and the personnel at Morton Publishing Company for the opportunity, encouragement, and support in this project.

About the Authors

Byron J. Adams completed his undergraduate degree in Zoology in 1993 from Brigham Young University with an emphasis in marine biology and his Ph.D. in Biological Sciences from the University of Nebraska in 1998. Following a short stint as a postdoctoral fellow at the University of California-Davis, Byron took his first faculty position at the University of Florida prior to returning to Brigham Young University. Byron's most recent projects involve fieldwork in Antarctica and research programs in biodiversity, evolution, and ecology.

John L. Crawley received his degree in Zoology from Brigham Young University in 1988. John has worked as a researcher for the National Forest Service and the Utah Division of Wildlife Resources, but now he devotes his time to observing nature, traveling, and taking photographs of wildlife, many of which have appeared in advertisements and publications for Delta Airlines, *National Geographic*, the Bureau of Land Management, the U.S. Forest Service, and Morton Publishing Company.

Byron on the plane making his way back from the Transantarctic Mountains heading for McMurdo Station.

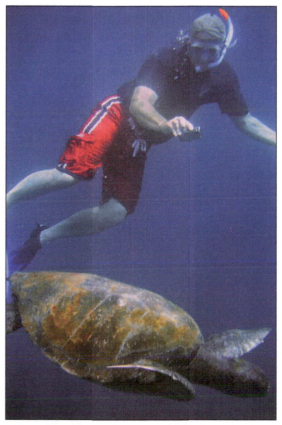

John snorkeling with green sea turtles in the Galapagos.

C O N T E N T S

UNIT 1

Introduction

• • • • • • • • •

In This Section

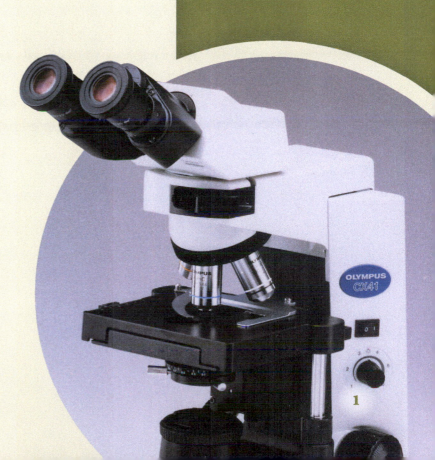

Exercise 1
Characteristics of Living Things

Objectives

At the completion of this exercise, students will be able to:

1 Define and apply characteristics of living things to common objects.

2 Demonstrate an increased awareness of the distinction between non-living and living things.

3 Apply commonly accepted characteristics of life to multiple phenomena and explore the boundaries of "life."

ⓘ Background Information

Biology is the study of life, and it may seem quite simple to distinguish between non-living and living things. However, in nature many things are more complex than they first appear. What is life? How can you distinguish living from non-living things? Scientists don't have a universally agreed upon definition of life, but most would refer to these seven commonly held characteristics of living things:

1. Homeostasis.
2. Response to stimuli.
3. Reproduction.
4. Evolution and Adaption.
5. Growth.
6. Organization (one or more cells).
7. Metabolism.

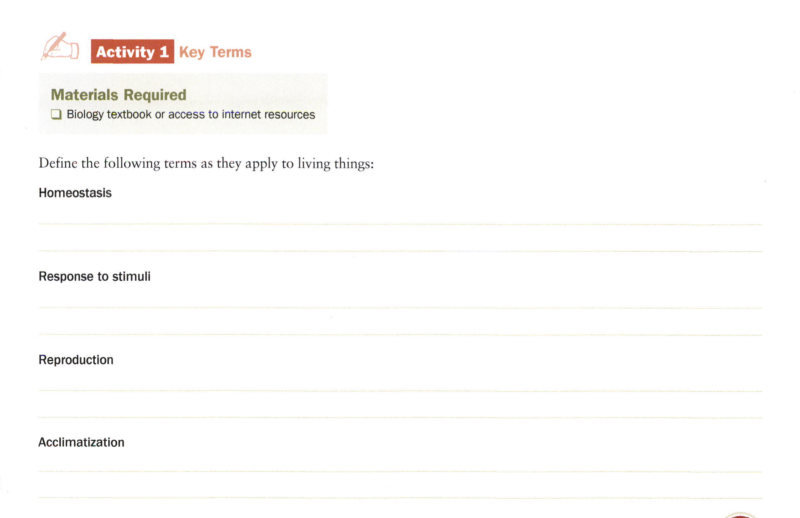

Activity 1 Key Terms

Materials Required
❑ Biology textbook or access to internet resources

Define the following terms as they apply to living things:

Homeostasis

_____

_____

Response to stimuli

_____

_____

Reproduction

_____

_____

Acclimatization

_____

_____

Adaptation

Evolution (*Note:* populations, not individuals, evolve.)

Growth

Organization (one or more cells)

Metabolism

Activity 2 Careful Observation of Non-Living and Living Things

Materials Required
☐ Biology textbook or access to internet resources

1 For each of the following examples, determine which of the seven characteristics of living things apply and check the relevant boxes. You may use the internet and other resources to determine all the characteristics.

2 Circle any of the seven characteristics of living things that each example is missing.

Fire

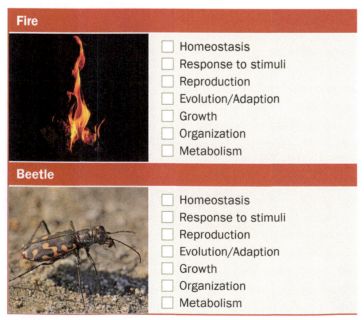

☐ Homeostasis
☐ Response to stimuli
☐ Reproduction
☐ Evolution/Adaption
☐ Growth
☐ Organization
☐ Metabolism

Beetle

☐ Homeostasis
☐ Response to stimuli
☐ Reproduction
☐ Evolution/Adaption
☐ Growth
☐ Organization
☐ Metabolism

Crystals

☐ Homeostasis
☐ Response to stimuli
☐ Reproduction
☐ Evolution/Adaption
☐ Growth
☐ Organization
☐ Metabolism

Stream

☐ Homeostasis
☐ Response to stimuli
☐ Reproduction
☐ Evolution/Adaption
☐ Growth
☐ Organization
☐ Metabolism

Mycoplasma		Tree	
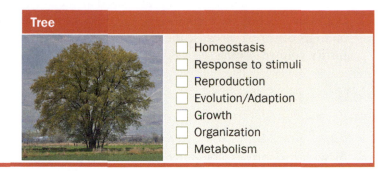	☐ Homeostasis ☐ Response to stimuli ☐ Reproduction ☐ Evolution/Adaption ☐ Growth ☐ Organization ☐ Metabolism	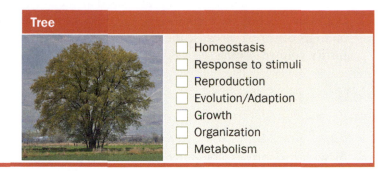	☐ Homeostasis ☐ Response to stimuli ☐ Reproduction ☐ Evolution/Adaption ☐ Growth ☐ Organization ☐ Metabolism

 Activity 3 Identifying Things That Blur the Line

Materials Required
❏ Biology textbook or access to internet resources

1 Using the internet and other resources, identify two things that always have all but one of the seven characteristics of living things, and identify two living things that could be confused for non-living things. Justify your choices.

_____
_____
_____
_____
_____
_____

 **Activity 4** Synthesis

Materials Required
No materials required for this activity

Now let's tie it all together.

1 Viruses can only replicate and make the products of their genes using the cellular machinery of other organisms. Using the characteristics from Activity 2, are they "alive"? Justify your reasoning.

_____
_____
_____

2 Scientists have engineered and synthesized artificial chromosomes and inserted them into brewer's yeast. The yeast expresses genes on the chromosomes, replicates, and passes new copies of the artificial, but fully functional chromosome on to its offspring. Because the genetic material of this organism was "made in the lab," is this organism "alive"? Justify your reasoning.

3 As with real viruses, computer viruses use the software and hardware of its host computer to replicate and spread to other computers. Are computer viruses "alive"? Justify your reasoning.

4 List the characteristics of each of the following examples and justify each as living or non-living.

a. Virus

b. Cut tree branch

c. Cut grapevine

d. Mule

5 Astrobiologists are interested in what life might be like elsewhere in the Universe, including life that might be based on completely different chemistries (inorganic versus organic) or information storage and transmission systems that are unlike anything known on Earth. Thus, astrobiologists adopt a broader definition of life: "A self-sustaining chemical system capable of Darwinian evolution." If you were to exclusively apply this definition to the specific cases in Activity 2 (Fire, Beetle, Crystals, Stream, Mycoplasma, Tree), which of these that you previously classified as "living" would be non-living according to the astrobiological definition of life? Which of the non-living things would be classified as alive?

Exercise 2
The Nature of Science

Objectives

At the completion of this exercise, students will be able to:

1 Define and apply key terms associated with the nature of science.

2 Compare and contrast the scientific enterprise with other ways of knowing.

3 Evaluate the attributes of some of the most common scientific and pseudoscientific topics.

4 Describe the difference between skepticism and denialism.

5 Distinguish between scientific and ethical, moral, or political questions.

6 Describe the impact of pseudoscience on health, politics, education, and ethics or religion.

ⓘ Background Information

Suppose you drop your pencil, and it falls to the floor. You might conclude that this occurrence is best explained by the theory of gravity. But how do you know that this explanation is correct? Is it because science has rigorously tested this theory? Are there other ways of knowing? What is the difference between science and pseudoscience? What is the difference between skepticism and denialism? What kinds of questions can and cannot be answered by science? In this exercise, you will explore these questions by applying your knowledge of these terms and definitions to various activities that commonly involve the scientific enterprise.

Activity 1 Key Terms

Materials Required
❏ Biology textbook or access to internet resources

Define the following terms:

Science

_____

_____

_____

Pseudoscience

_____

_____

_____

Falsification

_____

_____

_____

Verification

Skepticism

Denialism

Probability

Replication

Bias

Confirmation bias

Activity 2 How Do We Know What We Know?

Materials Required
❑ Biology textbook or access to internet resources

1 Science is a terrific way to "get to the bottom of something," but it is not the only way to learn about the world around us. In the space provided, compare and contrast each of these ways of knowing with scientific enterprise. How does each of these ways of knowing differ from science?

  a. Sensory perception

   Compare: _____

   _____

   Contrast: _____

   _____

  b. Logic/reason

   Compare: _____

   _____

   Contrast: _____

   _____

  c. Faith

   Compare: _____

   _____

   Contrast: _____

   _____

  d. Authority

   Compare: _____

   _____

   Contrast: _____

   _____

  e. Imagination

   Compare: _____

   _____

   Contrast: _____

   _____

2 Can you list some other ways of knowing?

Activity 3 **Is Astrology Scientific?**

Materials Required
☐ Biology textbook or access to internet resources

Astrologers claim that the positions of the sun, moon, and planets at the moment of birth can be used to determine a person's general personality traits and tendencies, as well as predict the major events and issues they are likely to encounter. Similarly, the Chinese zodiac consists of twelve animal signs representing twelve different types of personality.

In this exercise you will perform a simple experiment to see how well the Chinese zodiac predicts the type of person you are. To do this you will randomize the different personality traits associated with each animal and year of the zodiac, choose the set of traits that best describes you, and then see if it matches the traits ascribed to your birth year by the Chinese zodiac. Consider the following before conducting the experiment:

Question: Do Chinese zodiac traits accurately associate with their respective birth dates?

Null Hypothesis: There is no correlation between birth year and personality traits.

Alternative Hypothesis: Personality traits associate with birth year.

Test/Experiment: Without knowing the year or animal of the Chinese zodiac, identify the one letter (a–l) of the list of traits with which you feel you best self-identify. Check the key (the correctly associated zodiac) on page 17 to see if you picked the correct one.

1 Read the list of traits below and choose the one that best describes your personality.

   a. I am hardworking, industrious, and shrewd. I am also charming and have a magnetic personality. It's said that I will never be poor, since I am good in business.

   b. I am dependable, patient, and methodical. I pay attention to the small things in life and am not the type of person to be noticed by superiors. I am poor at communication.

   c. I am rebellious, passionate, and generous. I am here to make a splash in whatever area I choose. I am tolerant, staunch, and vigilant. In ancient times, my zodiac sign was associated with emperors and kings.

   d. I am gracious, kind, and sensitive. I like to express myself through art. I have an excellent memory and like to make other people laugh. I am good at creating fun and excitement in my life and the lives of others.

   e. I am proud, noble, and passionate. I am stately, magnanimous, and loyal to my friends. I tend toward perfectionism and am goal oriented.

   f. I am a deep thinker and tend toward the mystical. I am intelligent but can be secretive. I am a good communicator and can put others at ease with just a word.

   g. I am warmhearted and interested in improving myself. I am the spotlight of my community and like to help others. I like to stay active and am kind to most people that I meet.

h. I have friends throughout many different social classes. I am tender, kind, and generous. I like to live a quiet life.

i. I am lively and active. I have a strong desire to be my own person and live in a way that allows me to be free of restraint. I have a genuine interest in helping others.

j. I am honest, bright, and talkative. I have a tendency to flip-flop on issues and can be passionate about many different things.

k. I am gentle, lovable, and kind. I am loyal and courageous under pressure. I like to stay comfortable and dislike overly loud people.

l. I am honest with those in my life. I am loyal and sincere, and I expect others to be the same.

2 Which trait letter above best fits your personality?

3 In which year were you born?

4 Using your birth date and the key on page 17, what is your Chinese zodiac sign?

5 Again, using the key at the end of this exercise, does the trait letter you chose in question 2 match what was assigned by your Chinese zodiac birth year?

6 What is the probability that your choices match by chance alone?

7 Why would a skeptic ask you to replicate the experiment?

8 What other aspects of the experiment could be subject to bias?

9 How could confirmation bias influence the outcome of this experiment?

Materials Required

❏ Biology textbook or access to internet resources

Science is a way of pursuing knowledge that involves testable predictions and explanations about the natural world. Pseudoscience is any idea put forth as scientific when it is not.

In general, scientific ideas:

1. Are based on empirical, measureable observation.
2. Are confirmed by empirical tests or the discovery of new facts.
3. Do not respect authority. They are impersonal and, therefore, testable by anyone.
4. Explain a range of empirical phenomena.
5. Are empirically testable (falsifiable), such as testing specific predictions that arise from a theory.
6. Productive; they lead investigators to new knowledge and understanding.
7. Are put forth not as absolute truth, but as a best estimate and subject to error and re-examination and reinterpretation.
8. Are self-correcting.

On the other hand, most pseudoscientific ideas:

1. Have vague, as opposed to explicit, claims.
2. Exaggerate their claim.
3. Have claims that are untestable.
4. Rely on confirmation (not falsification).
5. Are not open to testing by others.
6. Do not result in new ideas, knowledge, or research programs.
7. Appeal to authority.
8. Use misleading language.

Most pseudoscientific claims are based on statements made by people professing to be experts or authorities in their fields. Rather than careful observation or empirical investigation, their claims are usually based only on evidence that confirms their existing notions (i.e., cherry-picking and confirmation bias) and thus cannot lead to new scientific discoveries. Science, on the other hand, not only leads to new discoveries but also corrects what we got wrong earlier. For example, in 1912 the skull of a modern human, the modified jawbone of an orangutan, and fossil chimpanzee teeth were packaged up as a composite and presented to the world as a "missing link" between apes and humans. At the outset, several scientists were skeptical of the claim. Within three years it was argued to be a fraud, and because the forgery was completely inconsistent with other finds, more and more aspersion was cast on its validity with each new find. Eventually (over 40 years later) the consilience of chemical tests and anatomical measurements conclusively demonstrated the counterfeit.

1 In the space provided, evaluate each topic by applying the characteristics of scientific ideas. Designate each as science or pseudoscience, and follow with your reasoning. For those characteristics you are not familiar with, use internet resources to learn more about them.

> **Note:** Use what you learned about science versus pseudoscience to identify internet resources that won't feed you a bunch of baloney!

    a. Ancient astronaut hypothesis

    b. Climate change denialism

    c. Chiropractic (Palmer School)

    d. Germ theory of disease

    e. Folk epidemiology of autism

    f. Homeopathy

    g. Magnet therapy

    h. Palm readings

i. Intelligent design

j. Cryptozoology

k. Theory of evolution

l. Theory of gravity

m. Holocaust denialism

n. Plate tectonics

o. Genetic theory of inheritance

p. Immunizations

q. Electromagnetic hypersensitivity

Activity 5 Skepticism versus Denialism

Materials Required
❏ Biology textbook or access to internet resources

Skepticism is a healthy, questioning attitude about a claim, fact, knowledge, or opinion. Denialism is when a person uses rhetorical tactics that are designed to give the impression that there is a legitimate debate among scientists when in fact there is none. Skepticism is often conflated with denialism at the interface between science, ethics, and politics. For example, it is not uncommon for people to confuse scientific questions with political questions. The difference between scientific questions and political questions is that scientific questions will always yield a correct or incorrect answer, whereas (morals and values aside) political questions do not. However, answers to political questions can be informed by science.

1 Identify each of the following questions as a scientific or political question:

   a. How do human emissions affect the climate?

   _____

   b. How much should we cut our emissions?

   _____

   c. Does thimerosal in vaccines harm children?

   _____

   d. What should we do about parents who refuse to vaccinate their children?

   _____

   e. Do mobile phone towers cause leukemia?

   _____

   f. How far away should phone towers be from schools?

   _____

   g. Are salmon stocks in the Pacific Northwest in decline?

   _____

   h. How many salmon can be sustainably harvested?

   _____

   i. Should politics be informed by science or pseudoscience? Defend your answer.

   _____

   _____

   _____

   _____

Activity 6 Synthesis

Materials Required

☐ Biology textbook or access to internet resources

Now let's tie it all together.

1 Identify and describe the implications of pseudoscience on the following topics:

 a. Politics

_____

_____

_____

_____

 b. Health

_____

_____

_____

_____

 c. Education

_____

_____

_____

_____

 d. Ethics/Religion

_____

_____

_____

_____

Astrology Activity Key (from Activity 3)

A. **The Rat:** (Birthdays in 1912, 1924, 1936, 1948, 1960, 1972, 1984, 1996, 2008) The Rat is hardworking, industrious, and shrewd. People born in the year of the Rat are also charming and have magnetic personalities. It's said that finding a poor Rat is rare, since they are good in business.

B. **The Ox:** (Birthdays in 1913, 1925, 1937, 1949, 1961, 1973, 1985, 1997, 2009) The Ox is dependable, patient, and methodical. Oxen pay attention to the small things in life and aren't the type of people to be noticed by superiors. They can be poor at communication.

C. **The Tiger:** (Birthdays in 1914, 1926, 1938, 1950, 1962, 1974, 1986, 1998, 2010) The Tiger is rebellious, passionate, and generous. A person born in the year of Tiger is here to make a splash in whatever area chosen. Tigers are tolerant, staunch, and vigilant. In ancient times, the sign of the Tiger was associated with emperors and kings.

D. **The Rabbit:** (Birthdays in 1915, 1927, 1939, 1951, 1963, 1975, 1987, 1999, 2011) Rabbits are gracious, kind, and sensitive. They like to express themselves through art. They have strong memories and like to make other people laugh. They are good at creating fun and excitement in their lives and the lives of others.

E. **The Dragon:** (Birthdays in 1916, 1928, 1940, 1952, 1964, 1976, 1988, 2000, 2012) People born in the year of the Dragon are proud, noble, and passionate. They are stately, magnanimous, and loyal to their friends. They can tend toward perfectionism and are goal oriented.

F. **The Snake:** (Birthdays in 1917, 1929, 1941, 1953, 1965, 1977, 1989, 2001, 2013) Snakes are deep thinkers and tend toward the mystical. They are intelligent but can be secretive. They are good communicators and can put others at ease with just a word.

G. **The Horse:** (Birthdays in 1918, 1930, 1942, 1954, 1966, 1978, 1990, 2002, 2014) Horses are warmhearted and interested in improving themselves. They like to stay active and are kind to most people that they meet. A horse is the spotlight of his or her community and likes to help others.

H. **The Goat:** (Birthdays in 1919, 1931, 1943, 1955, 1967, 1979, 1991, 2003, 2015) Goats have friends throughout many different social classes. People born in a Goat year are tender, kind, and generous. They like to live quiet lives.

I. **The Monkey:** (Birthdays in 1920, 1932, 1944, 1956, 1968, 1980, 1992, 2004, 2016) People born in Monkey years are lively and active. A Monkey has a strong desire to be his or her own person and live in a way that allows him or her to be free of restraint. Monkeys have a genuine interest in helping others.

J. **The Rooster:** (Birthdays in 1921, 1933, 1945, 1957, 1969, 1981, 1993, 2005, 2017) Roosters are honest, bright, and talkative. They have a tendency to flip-flop on issues and can be passionate about many different things.

K. **The Dog:** (Birthdays in 1922, 1934, 1946, 1958, 1970, 1982, 1994, 2006, 2018) Dogs are gentle, lovable, and kind. They are loyal and courageous under pressure. They like to stay comfortable and dislike overly loud people.

L. **The Pig:** (Birthdays in 1923, 1935, 1947, 1959, 1971, 1983, 1995, 2007, 2019) The year of the Pig gives birth to people who are honest with those in their lives. They are loyal and sincere, and they expect others to be the same.

Exercise 3
Scientific Method

Objectives

At the completion of this exercise, students will be able to:

1 Define and apply key terms related to scientific methods.

2 Integrate careful observation with previous knowledge in order to design a controlled experiment that can falsify hypotheses.

3 Describe multiple ways of testing hypotheses including controlled experiments, observational studies (or natural experiments), comparative methods, and computer simulations.

4 Describe the purpose of experimental controls.

5 Develop basic technical skills required to effectively analyze, interpret, and synthesize experimental data.

6 Describe basic principles and concepts associated with scientific (experimental) methods.

7 Plan and perform a simple experiment, draw conclusions, and effectively communicate the conclusions that logically follow from the results.

ⓘ Background Information

What is commonly referred to as "the scientific method" is probably more accurately described as any technique or approach that generates scientific knowledge. To be considered "scientific," the methods of inquiry must be based on empirical and measurable evidence subject to principles of logic and reason. For the natural sciences, this means "scientific" findings are those that follow from careful observations, measurements, and experiments. Scientific experiments are, specifically, those that can falsify hypotheses. Testable hypotheses serve as explanations for all observable phenomena in the natural world.

   The process of generating scientific knowledge involves making hypotheses (conjectures) that can be falsified, making predictions of logical consequences based on these hypotheses, and then performing experiments or recording observations that could falsify or otherwise determine whether an original hypothesis was correct. Although the scientific method is often presented as a chronological series of steps, these are more commonly considered general principles for most working scientists.

Formulation of a question: Why do cockroaches run away when I turn the lights on? How could I design an environmentally friendly pesticide? These two types of questions (the latter being open-ended) are the first steps in applying the scientific method to scientific research. At this point, most scientists inform the formulation of their question by considering all the findings of previous work on the subject done by other scientists.

Hypothesis: This is an explanation (conjecture) that you come up with that is based on all the knowledge and evidence you currently have at hand. Statistical hypotheses include consideration of both null and alternative hypotheses. The null hypothesis is that your explanation is not any different from chance alone. The alternative hypothesis is that your explanation is better than random. The most important consideration of a hypothesis is that it must be falsifiable, meaning that at least one possible outcome of your experiment is that the results conflict with the predictions that you deduced from your hypothesis. If the results could not possibly conflict with your predictions, then your hypothesis is not logically testable.

Prediction: This step involves determining the logical consequences of your hypothesis. Ideally your prediction must distinguish the hypothesis from likely alternatives. If your hypothesis for why cockroaches run away when you turn the lights on is that they are afraid of the light, then you might predict that cockroaches will run for cover every time the lights are turned on, but stop when the lights are turned off.

Testing: This is the step where you check to see if what happens in the real world is the same as what is predicted by your hypothesis. You can test your hypotheses by performing experiments. Does the outcome of your experiment agree with the predictions derived from your hypothesis? If yes, then you can have greater confidence in your hypothesis.

Controls: To minimize experimental error, scientists use controls. For example, suppose you hypothesize that hummingbirds are attracted to sugar water. To test your hypothesis, you mix up a bunch of sugar in water and put it in your feeder, and sure enough, the hummingbirds come to your feeder. How do you know it is the sugar that they are attracted to and not the water? Maybe they are just thirsty? To get around this problem you could implement an experimental control. In this case, the hummingbirds could be exposed to two feeders—one with water only and the other with a sugar-water solution. The water-only solution serves as your control or your placebo. Now you can test if significantly more birds come to the sugar water than the water-only feeder.

Analysis: What are the results of your experiment? Experimental analysis involves comparing the predictions that follow from your hypothesis compared to the null hypothesis and determining which hypothesis better explains your data.

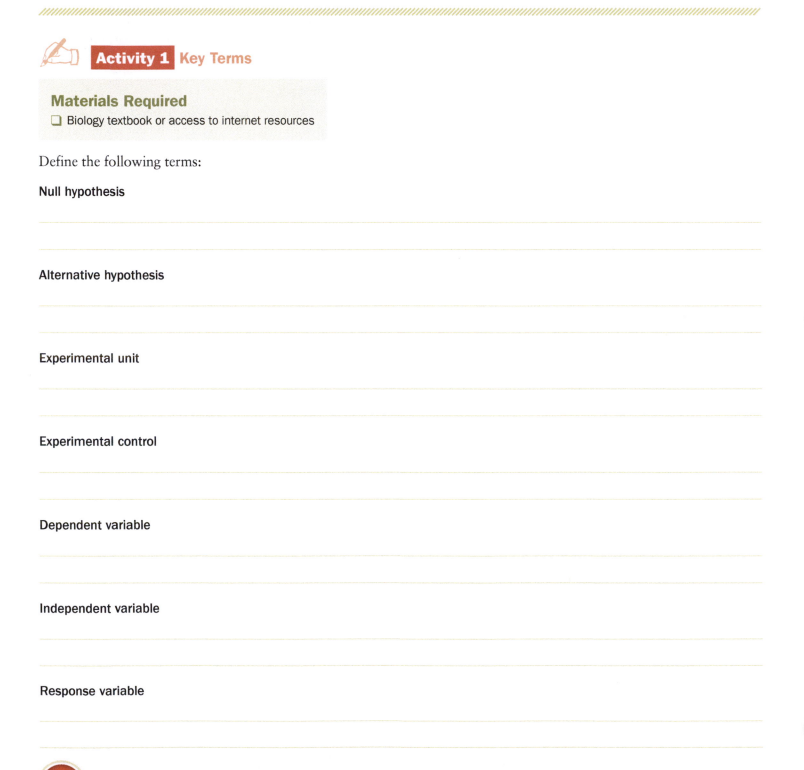

Activity 1 Key Terms

Materials Required
❑ Biology textbook or access to internet resources

Define the following terms:

Null hypothesis

_____

_____

Alternative hypothesis

_____

_____

Experimental unit

_____

_____

Experimental control

_____

_____

Dependent variable

_____

_____

Independent variable

_____

_____

Response variable

_____

_____

Taxis

_____

_____

Kinesis

_____

_____

Activity 2 **Careful Observation of Animal Behavior**

Materials Required
- ❑ Biology textbook or access to internet resources
- ❑ 2 identical cake pans (or similar containers)
- ❑ Paper towel, or paper napkins
- ❑ 10 individual animals of the same species

1 Collect ten harmless, crawling, non-flying invertebrate animals of the same species that are readily available to you. These are organisms that live all around us, such as pill bugs (woodlice), grasshoppers, beetles, cockroaches, caterpillars, snails, slugs, and earthworms. If winter or environmental conditions prevent you from collecting ten individuals of the same species (indoor or outdoor), you can find suitable alternatives such as insect larvae (wax worms or mealworms), crickets, or earthworms at a local pet store or fishing tackle bait shop.

Note: Avoid using biting or stinging organisms, and don't use vertebrate animals at all.

2 Answer the following questions about the natural habitat of your collected species:

a. Where do they normally live?

_____

b. Do they normally live above or below ground?

_____

c. What kind of temperatures do they normally experience?

_____

d. Is their habitat typically wet or dry?

_____

e. Do they live in lighted or dark environments?

_____

3 Place five of your experimental animals in each of your observation arenas (cake pans). Don't let them escape!

4 Observe the animals for five minutes.

5 Answer the following questions about the external morphology and movement of your animal:

a. What does your animal look like?

_____

_____

_____

b. Sketch a picture of your animal in the space provided. Identify the main features of its external morphology (its form and structure). If necessary, use your textbook or internet resources as a reference.

c. How does your animal move? Describe how your animal moves, including which muscles are necessary. Use your textbook and internet resources as a reference to identify which muscles and appendages are involved in locomotion.

_____

_____

_____

6 Taxis is an innate behavioral response by an organism to a directional stimulus or gradient. Kinesis is a non-directional change in activity in response to a stimulus. Answer the following questions about kinesis and taxis:

a. Compare and contrast kinesis and taxis.

_____

_____

_____

b. What is positive phototaxis?

_____

_____

c. What is negative phototaxis?

_____

_____

Activity 3 Generating Testable and Falsifiable Hypotheses

Materials Required
- ❑ 10 individual animals of the same species
- ❑ 2 identical cake pans (or similar containers)
- ❑ Paper towel, or paper napkins
- ❑ Water
- ❑ Flashlight

1 Now you will generate a testable hypothesis. Based on your observations about their natural environment, predict whether or not your animals will prefer a lighted or dark habitat.

a. Null hypothesis: _____

_____

b. Alternative hypothesis: _____

_____

2 Now you will test a falsifiable hypothesis. Design an experiment that could lead you to reject the null hypothesis using your animals and the materials listed at the beginning of this activity.

a. Describe the experimental group.

_____

b. Describe the control group.

_____

c. What is the purpose of the control group?

_____

d. Describe the experimental units in your experiment. For example, are you measuring the responses of each individual animal or of populations of animals?

_____

e. What is the treatment in your experiment?

_____

_____

f. What is the response variable in your experiment (i.e., what are you measuring or counting)?

_____

_____

3 In the space provided, sketch your experimental setup.

4 Describe your experimental results.

5 Draw a graphical representation of your experimental results.

6 What conclusions can you draw from your experimental results? Describe the conclusions of your research in a single, concise sentence.

7 How might sample size influence your results?

8 How does the number of trials (replication) influence your results?

9 List the caveats and potential sources of unaccounted errors in the design and execution of your experiment.

Activity 4 Synthesis

Materials Required
❏ Biology textbook or access to internet resources

Now let's tie it all together.

1 Identify the parts of the scientific method in a classic "historic" experiment, such as Francesco Redi's test of spontaneous generation.

Exercise 4
Experimental, Observational, and Theoretical Science

Objectives

At the completion of this exercise, students will be able to:

1 Define and apply key terms associated with scientific processes.

2 Explain the difference between experimental, observational, and theoretical science.

3 Design an experiment using key elements of scientific methods.

4 Demonstrate figure literacy by interpreting graphical scientific data.

5 Infer logical conclusions from scientific data.

ⓘ Background Information

Although students are frequently taught "the scientific method," there are numerous methods scientists use to make discoveries, solve problems, and explore how the natural world works (as we scientists say, "get to the bottom of things"). While controlled experiments are extremely powerful, they are not always feasible. For example, if we want to study how hurricanes work, some of our most informative inferences will come from careful observation of how they form in nature. Experimental and observational scientists use statistical inference in order to determine if the differences they see among their observations are significant. Other experiments can be conceived in our minds but may be very difficult to test in practice, such as some of the theoretical ideas Albert Einstein came up with that involve the effects of time and mass when traveling at the speed of light. Although experimental, observational, and theoretical approaches differ in terms of the types of data that are collected and how they are analyzed, each provides important avenues in furthering our understanding of how the world works.

✏️ Activity 1 Key Terms

Materials Required
☐ Biology textbook or access to internet resources

Define the following terms:

Experimental science

_____

_____

Observational science

_____

_____

Theoretical science

_____

_____

Experimental unit

Response variable

Independent variable

Dependent variable

Null hypothesis

Control

Treatment

Placebo

Replication (scientific method)

Sample size

Alternative hypothesis

Standard deviation

Standard error of the mean

P-value

"Different" versus "significantly different"

Linear regression line

Activity 2 **Efficacy of a New Facial Cream**

Materials Required
No materials required for this activity

With the help of the lead singer of a hugely popular rock band, a drug company is successfully marketing a new facial cream that claims to treat acne better than any other existing product. You are charged with determining the efficacy of this new drug. You recruit 50 people to help you with your study. You assign 25 of them to a placebo group, and 25 to the treatment (medication) group. After a 12-week period you compare the results (number of facial lesions before, and after) of each group.

1 What is your null hypothesis?

2 What is your alternative hypothesis?

3 What is the experimental unit in your study?

4 What is the purpose of the placebo group?

5 What is the dependent variable in your study?

6 What is the independent variable in your study?

7 What steps can you take to determine whether the new drug works significantly better than the placebo?

8 Suppose the participants in the study show significantly fewer lesions than the placebo group after 12 weeks. Can you conclude that it is the best treatment for acne available? Why or why not?

9 How would you modify the experiment to test the company's claim that their drug is the best treatment option available?

10 What are some confounding factors that could nullify the conclusions drawn from this experiment?

11 What changes to the experimental design could be made to address these confounding factors?

12 This is an example of what kind of science?

Activity 3 Antarctic Worms and Climate Change

Materials Required
No materials required for this activity

Dr. Scott Nema has been studying Antarctic worms in the McMurdo Dry Valleys for the past 25 austral summers, collecting information on their abundance, spatial distribution, survival strategies, and trophic habits. In the same location his colleagues have been taking careful temperature measurements. With the help of his colleagues, Dr. Nema has noticed that soil temperatures became significantly cooler between 1986 and 2000 (Fig. 4.1), and nematode abundance decreased significantly over the same time period (Fig. 4.2). Dr. Nema suspects that in response to climate change, the rate and amount of carbon cycled in Antarctic soil ecosystems will decrease significantly.[1]

[1] Doran, P. T., J. C. Priscu, W. B. Lyons, J. E. Walsh, A. G. Fountain, D. M. McKnight, D. L. Moorhead, R. A. Virginia, D. H. Wall, G. D. Clow, C. H. Fritsen, C. P. McKay, and A. N. Parsons. 2002. Antarctic climate cooling and terrestrial ecosystem response. Nature 415:517-520. http://www.nytimes.com/2006/07/27/opinion/27doran.html?_r=0.

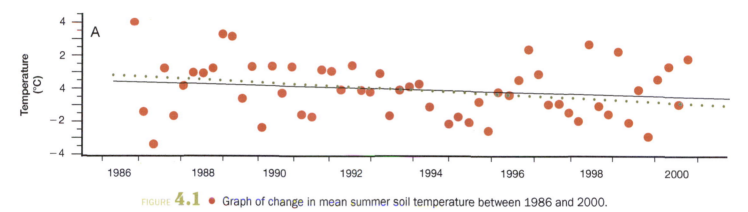

FIGURE **4.1** • Graph of change in mean summer soil temperature between 1986 and 2000.

1 This is an example of what kind of science?

2 What are the dependent and independent variables in Figure 4.1?

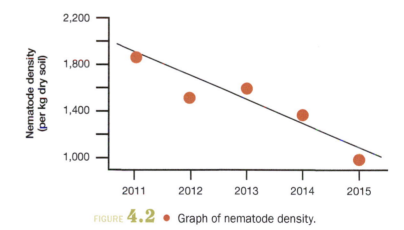

FIGURE **4.2** • Graph of nematode density.

3 What are the dependent and independent variables in Figure 4.2?

4 How can you determine if the decline in nematode abundance and the decrease in soil temperature are significant?

_____

_____

_____

5 What conclusions about climate can you draw from these observations?

_____

_____

_____

6 Because he knows how much bacteria are eaten by these Antarctic nematodes, Dr. Nema can calculate their contribution to the carbon cycle. From the results of this analysis, can you conclude that climate change is driving changes in ecosystem functioning (carbon cycling)? Why or why not?

_____

_____

_____

_____

 **Activity 4** **Your Own Design**

Materials Required
No materials required for this activity

1 Design your own experimental approach to "getting to the bottom of something." Include in your experimental design the following:

a. null hypothesis

b. alternative hypothesis

c. experimental unit

d. response variable

e. independent variable

f. dependent variable

g. control

h. treatment

i. sample size

j. replication (number of trials)

2 Assume the results of your experiment favor the alternative hypothesis.

    a. Make a table that accurately displays your data.

    b. Carefully construct a graph that accurately displays your data and clearly shows support for the alternative hypothesis.

    c. Draw a graph of your imaginary results.

    d. Craft a one-sentence conclusion based on your results.

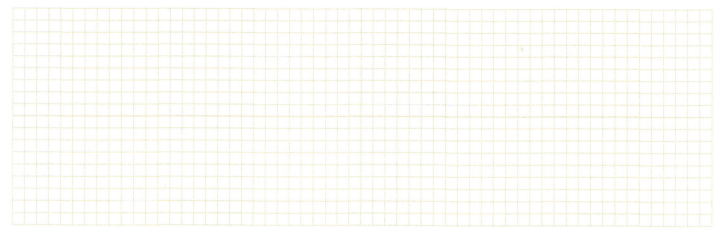

Conclusion: _____

_____

_____

3 Design your own observational approach to getting to the bottom of something. Include in your experimental design the following: null hypothesis, alternative hypothesis, experimental unit, response variable, independent variable, dependent variable, control, and treatment. Draw a graph of your imaginary results, and finish with a one-sentence conclusion based on your results.

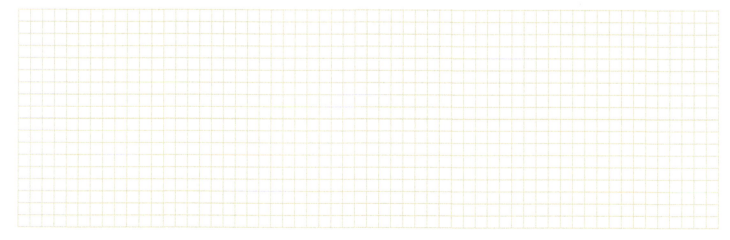

Conclusion: _____

_____

_____

 **Activity 5** Synthesis

Materials Required
No materials required for this activity

Now let's tie it all together.

1 Does one type of science produce answers that are more certain than the other? Why or why not?

2 Why is it important to make the distinction between "different" and "statistically significant"?

3 What effect does sample size and the number of replicates have on statistical significance?

4 When it comes to the interpretation of scientific data, why is it important to distinguish between what scientists are in control of, and what they are not?

5 What is the difference between the "results" of an experiment and the "conclusions" of an experiment?

Chemistry

In This Section

Exercise 5
Atomic Structure: Subatomic Particles and Electron Shells

Objectives

At the completion of this exercise, students will be able to:

1 Define and apply key terms associated with the atomic theory of matter.

2 Describe atomic structure by drawing, labeling, and identifying atoms.

3 Discuss the relationship between electrons and photons, and the role of atoms in electron (energy) transfer.

ⓘ Background Information

Atoms are the smallest units of matter that make up chemical elements. An element is a pure substance that contains only one kind of atom. Living things are mostly composed of six elements: carbon (C), hydrogen (H), nitrogen (N), oxygen (O), phosphorus (p), and sulfur (S). The chemical properties of these elements, such as the number and position of their electrons, determine the types of interactions they have with themselves and other elements. Living things need energy, and energy is carried by electrons.

Activity 1 Key Terms

Materials Required
❏ Biology textbook or access to internet resources

Define the following terms:

Atom

_____

_____

Electron

_____

_____

Neutron

_____

_____

Noble gas

_____

_____

_____

Electron shell

_____

_____

Atomic weight

_____

_____

Valence electron

_____

_____

Octet rule

_____

_____

Photon

_____

_____

Activity 2 Understanding Atomic Structure

Materials Required
❑ Biology textbook or access to internet resources

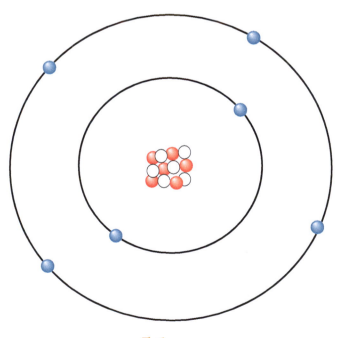

FIGURE **5.1** ● Sample atom.

1 Which atom is shown in Figure 5.1?

_____

2 How many electron shells does it have?

_____

3 How many electrons can it share?

_____

4 Draw and label a phosphorus atom (P).

5 How much does a neutron weigh?

_____

6 How much does an electron weigh?

_____

7 How much does a proton weigh?

_____

Activity 3 Synthesis

Materials Required
No materials required for this activity

Now let's tie it all together.

1 Why is all life on Earth carbon-based? To help you answer this question, consider the following:

 a. Is carbon abundant or rare in the universe?

 _____

 b. Is it a relatively large or small atom?

 _____

 c. Is it a relatively heavy or light atom?

 _____

 d. How many valence electrons does it have?

 _____

 e. Given the atomic properties of carbon, why do you think that most scientists believe that if life exists elsewhere in the universe, it will also be carbon-based?

 _____

 _____

 _____

2 How is the position of an electron related to the amount of energy it contains?

_____

_____

3 Compare and contrast an electron and a proton.

_____

_____

4 A phosphor (not to be confused with elemental phosphorus) is a luminescent substance that emits light. These are commonly found in glow-in-the-dark toys. Glow-in-the-dark toys can be "charged" by exposing them to light, which excites the electrons in the phosphor and causes them to emit light. In the space provided, draw a cartoon of an oxygen atom in the phosphorescent toy. Show the movement of electrons that occurs when the toy is "charging" and when it is "glowing."

Exercise 6
Chemical Bonding: Covalent, Ionic, and Hydrogen Bonds

Objectives

At the completion of this exercise, students will be able to:

1 Define and apply key terms associated with chemical bonding.
2 Describe the properties of covalent and ionic bonds in the context of their electronegativity.
3 Use an atom's electronegativity to predict the types of bonds that it will form.
4 Describe the structure of covalent and hydrogen bonds.
5 Distinguish between polar and nonpolar covalent bonds.
6 Describe fundamental properties of water and their importance to life on Earth.

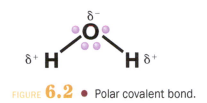

FIGURE **6.1** ● Nonpolar covalent bond.

FIGURE **6.2** ● Polar covalent bond.

ⓘ Background Information

Chemical bonds are the attractions between atoms that cause them to join together into substances of two or more atoms. The behavior of electrons determines whether a chemical bond will form and what shape the bond will have. The strength of these bonds varies. Atoms that share or transfer electrons have strong chemical bonds. Electrons that are shared equally are nonpolar (Fig. 6.1); electrons that are not shared equally result in partial charges and are polar (Fig. 6.2). Electrostatic attractions between the charged regions of polar molecules produce weak bonds (Fig. 6.3).

Electronegativity is the tendency of an atom to attract electrons. The higher an atom's electronegativity, the more strongly it attracts electrons.

Compared to most other molecules, water has some exceptional properties that make it fundamental to life on Earth. Unlike most substances, water expands when it freezes. Because many substances dissolve in water, it is commonly known as "the universal solvent." Its high specific heat capacity and heat of vaporization are due to the extensive hydrogen bonding between its polar molecules. All three states (liquid, solid, and gas) are found naturally in Earth. The human body is up to 78 percent water.

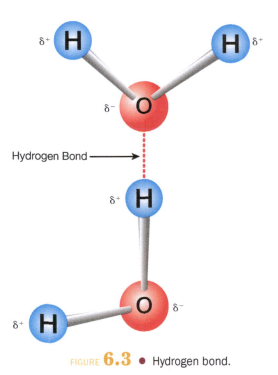

Hydrogen Bond →

FIGURE **6.3** ● Hydrogen bond.

 **Activity 1** Key Terms

Define the following terms:

Chemical bond

_____

_____

Covalent bonds

_____

_____

Polarity

_____

_____

Ionic bonds

_____

_____

Hydrogen bonds

_____

_____

Hydrophobic

_____

_____

Hydrophilic

_____

_____

Amphipathic

_____

_____

Electronegativity

_____

_____

Partial charge (net atomic charge)

_____

_____

Specific heat capacity

_____

_____

Heat of vaporization

_____

_____

Cohesion

_____

_____

Adhesion

_____

_____

Capillary action

_____

_____

Density of water and ice

_____

_____

Electronegativity Table of the Elements

1	2	3	4	5	6	7	8	9	10	11	12	13	14	15	16	17	18
H 2.1																	He 0
Li .98	Be 1.57											B 2.04	C 2.55	N 3.04	O 3.44	F 3.98	Ne 0
Na .93	Mg 1.31											Al 1.61	Si 1.9	P 2.19	S 2.58	Cl 3.16	Ar 0
K .82	Ca 1	Sc 1.36	Ti 1.54	V 1.63	Cr 1.66	Mn 1.55	Fe 1.83	Co 1.88	Ni 1.91	Cu 1.9	Zn 1.65	Ga 1.81	Ge 2.01	As 2.18	Se 2.55	Br 2.96	Kr 0
Rb .82	Sr .95	Y 1.22	Zr 1.33	Nb 1.6	Mo 2.16	Tc 1.9	Ru 2.2	Rh 2.28	Pd 2.2	Ag 1.93	Cd 1.69	In 1.78	Sn 1.96	Sb 2.05	Te 2.1	I 2.66	Xe 2.6
Cs .79	Ba .89	La 1.1	Hf 1.3	Ta 1.5	W 2.36	Re 1.9	Os 2.2	Ir 2.2	Pt 2.28	Au 2.54	Hg	Tl 2.04	Pb 2.33	Bi 2.02	Po 2	At 2.2	Rn 0
Fr .7	Ra .89	Ac 1.1	Rf	Db	Sg	Bh	Hs	Mt	Uun	Uuu	Uub						

Lanthanum

Ce 1.12	Pr 1.13	Nd 1.14	Pm 1.13	Sm 1.17	Eu 1.2	Gd 1.2	Tb 1.1	Dy 1.22	Ho 1.23	Er 1.24	Tm 1.25	Yb 1.1	Lu 1.27

Actinium

Th 1.3	Pa 1.5	U 1.38	Np 1.36	Pu 1.28	Am 1.3	Cm 1.3	Bk 1.3	Cf 1.3	Es 1.3	Fm 1.3	Md 1.3	No 1.3	Lr

Legend:
- 0–.66
- .66–1
- 1–1.33
- 1.33–1.66
- 1.66–2
- 2–2.33
- 2.33–2.66
- 2.66–
- White = No Data

FIGURE **6.4** • Electronegativity chart.

Activity 2 Electronegativity

Materials Required
No materials required for this activity

1 Using the electronegativity values shown in Figure 6.4 and Table 6.1 (which classifies bond types by their electronegativity), determine the bond types of the following three bonds. For each bond, perform the following steps:

Step 1: Determine the electronegativity of each element.

Step 2: Subtract the electronegativities and find where they lie on the table provided.

TABLE **6.1** Classification of Bond Types by Differences in Electronegativity between the Atoms Forming the Bond

0.0–0.3	Nonpolar covalent bond
0.4–1.9	Polar covalent bond
2.0–3.3	Ionic bond

a. Potassium (K) and chlorine (Cl)

What type of bond is this?

_____

b. Sulfur (S) and fluorine (F)

What type of bond is this?

_____

c. Iodine (I) and chlorine (Cl)

What type of bond is this?

_____

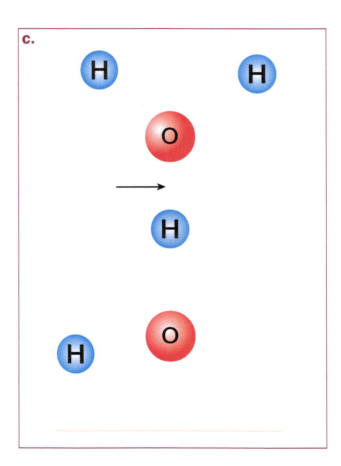

Activity 3 The Structure of Chemical Bonds

Materials Required

No materials required for this activity

1 Label each of the figures below. Be sure to include the bond, the type of bond (covalent polar, covalent nonpolar, hydrogen), electrons, and atomic charges.

a.

H H

b.

O

H   H

2 Can a hydrogen bond form between two atoms that share a nonpolar covalent bond? Why or why not?

Activity 4 Surface Tension

Materials Required
- ❏ Flat desktop or table
- ❏ Cup of water
- ❏ Small paper clips
- ❏ Coarse-ground black pepper
- ❏ Dish soap
- ❏ Paper towel

1 Considering that the paper clip is made of a metal that is denser than water, what do you predict will happen when you drop it in the water?

2 Drop the paper clip into the water. What do you observe?

3 Set a dry paper clip on a small piece of paper towel, and carefully lay the two flat on the surface of the water. What do you observe?

4 Instead of dropping the paper clip in the cup, gently lay it flat on the surface of the water. What do you observe?

5 Sprinkle the black pepper on the surface of the water. What do you observe?

6 Place a drop of soap in the water. What do you observe?

7 Soap looks like a triglyceride (see Fig. 6.6, p. 52). How do the soap molecules interact with the water molecules?

8 What effect does this have on surface tension?

Activity 5 Synthesis

Materials Required

No materials required for this activity

Now let's tie it all together.

Chemical Bonding

1 Draw a sodium atom (Na) and a chlorine atom (Cl) side-by-side, including the correct number of electrons in each of the orbitals.

a. How do these atoms appear when bonded to form sodium chloride (table salt; NaCl)?

b. What kind of bond is formed between sodium and chloride to form NaCl? Provide an explanation for your answer.

2 Draw a representation of how these molecules appear in the solution with water.

a. What kinds of bonds are formed between water and sodium? Provide an explanation for your answer.

_____

_____

_____

b. What kinds of bonds are formed between water and chlorine? Provide an explanation for your answer.

_____

_____

_____

3 Figure 6.5 is the chemical formula for a simple alcohol (n-propyl alcohol; 1-propanol).

$$CH_3 - CH_2 - CH_2 - OH$$

FIGURE **6.5** ● Chemical formula for a simple alcohol.

It can also be represented as ╱╲╱OH where the lines represent hydrogen-saturated carbon atoms with a hydroxyl (–OH) group on the end.

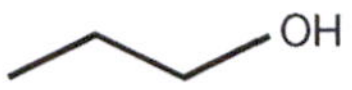

Draw a representation of how this molecule will appear in a solution of water.

4 Figure 6.6 is a diagram of a triglyceride: the combination of an alcohol (in this case glycerol) and three fatty acids. The hydroxyl groups of the glycerol join the carboxyl groups of the three fatty acids to form a triglyceride. In the space provided, draw a representation of how these molecules appear in the solution with water.

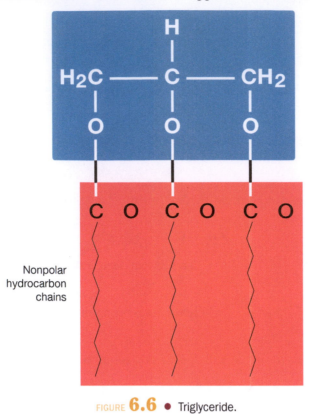

FIGURE **6.6** • Triglyceride.

Biological Applications of Chemical Bonding: Water

5 Water has high specific heat. What is the benefit of high specific heat to living organisms?

_____

_____

_____

6 Water has a high heat of vaporization. What is the benefit of high heat of vaporization to living organisms?

_____

_____

_____

7 Water is strongly cohesive. What is the benefit of cohesion to living organisms?

_____

_____

_____

8 Given that water expands when it is made into a solid, how is this important to life on earth? How is this costly to living organisms?

_____

_____

_____

9 Seattle, Washington is quite a bit farther north (higher in latitude) than Susanville, California. Because Seattle is closer to the North Pole, you might think it will be much colder in the winter. However, Seattle's winter climate is actually quite mild, with very little snow if any. Susanville, California, on the other hand, can experience extremely cold temperatures and lots of snow for much of the winter. Which properties of water might affect this, and how?

_____

_____

_____

10 One of the reasons humans are better endurance runners than any other mammal is because we have so many sweat glands. When the sweat (water) on your skin evaporates, energy from the sun (heat) breaks the hydrogen bonds between the water molecules, pulling heat away from your body and lowering your body temperature. Staying cool allows us to keep running when other mammals have to look for shade. What property of water helps us regulate body temperature by sweating?

How is this property important at the:

a. Cellular level?

b. Organismal level?

c. Ecosystem level?

11 Ice floats in water. What property of water is illustrated in this image of an iceberg?

How is this property important at the:

a. Cellular level?

b. Organismal level?

c. Ecosystem level?

12 This six-spotted fishing spider sits on top of the water. What property of water is illustrated in this image?

How is this property important at the:

a. Cellular level?

_____

_____

b. Organismal level?

_____

_____

c. Ecosystem level?

_____

_____

13 These polar bears huddle together to stay warm. What property of water is illustrated in this image?

How is this property important at the:

a. Cellular level?

_____

_____

b. Organismal level?

_____

_____

c. Ecosystem level?

_____

_____

Exercise 7
Biologically Important Compounds and Molecules

Objectives

At the completion of this exercise, students will be able to:

1 Define and apply key terms related to biologically important compounds and molecules.

2 Discuss principles and applications of hydrophobic, hydrophilic, amphipathic molecules, and solubility.

3 Apply an understanding of polar and nonpolar molecules to ecological remediation.

ⓘ Background Information

Biological membranes are made of phospholipid bilayers— a thin membrane made of two layers of lipid molecules that form a continuous barrier around cells (Fig. 7.1).

Phospholipids are amphipathic, possessing a polar, hydrophilic head (water-loving, reacts readily with water) and a nonpolar, hydrophobic (water-fearing) tail. They are comprised of two fatty acids and a phosphate compound bound to glycerol. The fat molecules that comprise the tails are hydrophobic, and the phosphate head has a negative charge, making it hydrophilic. When in liquid water, the molecules spontaneously arrange themselves into a two-layered sheet (bilayer). The tails of each molecule point toward the center of the two-layered sheet, where there is no water, while the heads point away from the sheet towards the inside or outside of the cell. Dishwashing detergent (Fig. 7.2) is much like a phospholipid. Each molecule is amphipathic, with a polar head and a nonpolar tail.

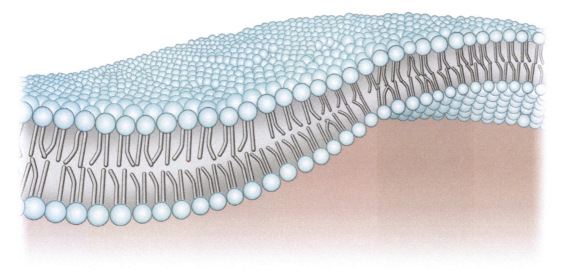

FIGURE **7.1** ● Phospholipid bilayer of a cell membrane.

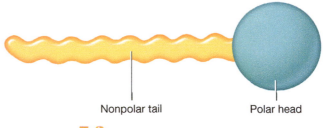

Nonpolar tail   Polar head

FIGURE **7.2** ● Molecule of dishwashing detergent.

Activity 1 Key Terms

Materials Required
❑ Biology textbook or access to internet resources

Define the following terms:

Molecular polarity

_____

_____

Hydrophobic

_____

_____

Hydrophilic

_____

_____

Amphipathic

_____

_____

Solubility

_____

_____

Activity 2 Observations in a Soda Bottle

Materials Required
- ❏ Access to internet resources
- ❏ Vegetable oil
- ❏ Water
- ❏ Clear plastic soft drink bottle with cap
- ❏ 1% dishwashing liquid detergent (Dawn works best)
- ❏ Food coloring

1 Fill the soda bottle half full of water and add a few drops of food coloring.

2 Pour 2 tablespoons of cooking oil into the soda bottle. Record your observations.

3 Screw the soda bottle lid on tight and shake vigorously. Record your observations.

4 Add dishwashing detergent and shake vigorously. Record your observations.

5 Use key terms from this activity to explain why oil and water do not mix.

6 Use key terms from this activity to explain why adding amphipathic molecules, via the soap, allows oil and water to mix.

7 Draw a sketch in the space provided that illustrates the principle of why dishwashing liquid removes greasy grime off of dinner plates. Include in your sketch water molecules, fat molecules, and soap molecules.

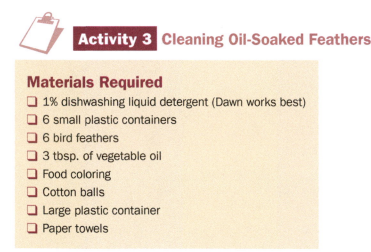

Activity 3 Cleaning Oil-Soaked Feathers

Materials Required
- ❏ 1% dishwashing liquid detergent (Dawn works best)
- ❏ 6 small plastic containers
- ❏ 6 bird feathers
- ❏ 3 tbsp. of vegetable oil
- ❏ Food coloring
- ❏ Cotton balls
- ❏ Large plastic container
- ❏ Paper towels

Oil spills have increased dramatically over the past two decades. Although natural petroleum seeps occasionally impact marine ecosystems, most spills are associated with drilling, shipping, or pipeline accidents.

Nearly all of the species in an ecosystem that experience a spill are affected, including marine invertebrates, birds, and mammals. Bird feathers are made of keratin, a tough protein similar in structure to your hair and fingernails. On a living bird, the feathers overlap each other like shingles on a roof to create a waterproof covering. However, when feathers come in contact with oil, they lose their waterproofing and insulating ability, resulting in hypothermia. Once oil is removed from feathers, the birds can preen, or straighten and clean, their feathers to restore their waterproofing and insulating abilities.

This activity is designed to show how dishwashing liquid can be used to clean the oil from bird feathers. Be mindful that the actual process is a highly coordinated and significantly more complex process. Oiled birds and other animals should not be treated or cleaned without proper facilities and experienced personnel. Because petroleum-based oils are toxic, we will use vegetable oil for this exercise to mimic the effects of crude oil on bird feathers.

1 Add the vegetable oil to one of the small plastic trays. Add a few drops of food coloring to the oil to simulate the dirty color of crude oil. Add three feathers to this solution and let them soak until you can see the feathers absorbing the oil.

2 Place three of the oil-soaked feathers in a small plastic tray containing only water.

3 Using cotton balls, gently scrub the feathers until the water becomes dirty with oil. When the water is dirty, move the feathers to another container, and so on until the feathers appear clean, or for a maximum of 5 minutes. Record your observations in the space provided.

4 Place three feathers in a small plastic tray containing 1% dishwashing solution. To make the 1% solution, at 1 tsp. of full-strength dish detergent to 2 cups of water (or 5 mL detergent to 236 mL water).

5 Use the cotton balls to gently scrub the feathers in the 1% dishwashing solution until the water becomes dirty with oil. When the water is dirty, move the feathers to another container, and so on, until the feathers appear clean, or for a maximum of 5 minutes.

6 When the feathers appear clean, rinse them with water. Record your observations below.

_____

_____

_____

_____

_____

_____

_____

_____

_____

7 Dispose of the feathers, water, and oil as indicated by your instructor.

8 Use key terms from this lesson to describe whether the water or the dishwashing solution worked most effectively.

_____

_____

_____

_____

_____

_____

9 Use key terms from this lesson to describe why the dishwashing liquid was effective in cleaning the feathers.

_____

_____

_____

_____

_____

_____

_____

10 Describe several problems that a professional or a volunteer may encounter in cleaning birds.

11 Even though the feathers may become clean, what other problems does the bird face?

Activity 4 Synthesis

Now let's tie it all together.

1 Besides cleaning oil from feathers, can you think of any other environmental applications of this principle?

_____

_____

_____

_____

2 Given the cost and time it takes to clean the feathers of oil-contaminated birds, is it worth it in the long run to mount rescue efforts in response to large oil spills? Justify your answer.

_____

_____

_____

_____

_____

_____

_____

Cell Biology

• • • • • • • • •

In This Section

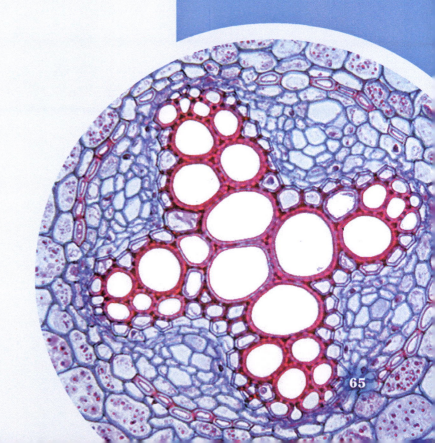

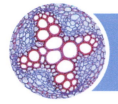

Exercise 8
Cell Membranes: Prokaryotic and Eukaryotic Cell Structures

Objectives

At the completion of this exercise, students will be able to:

1 Define and apply key terms associated with cell membranes and organelles.

2 Demonstrate knowledge and understanding of the difference between prokaryotic and eukaryotic cell structures.

3 Conduct experiments with cell membranes and demonstrate their importance to cellular life.

ⓘ Background Information

Cells are the fundamental units of life. All living organisms are composed of cells, and all cells come from preexisting cells. The universal characteristic of living cells is a selectively permeable membrane. The membrane allows cells to sequester certain molecules while preventing others from entering and facilitates communication and binding with its environment and other cells.

An important component of cell membranes is the phospholipid bilayer, composed of two fatty acids and a phosphate compound bound to glycerol. The phosphate group has a negative charge, making that part of the molecule hydrophilic. Similarly, bubbles are composed of lipid molecules and water and provide an excellent model for studying membranes. The lipid molecules are amphipathic, possessing a hydrophilic head (positive charge) and a hydrophobic tail. The bubbles form when the lipid molecules interact with air and water, creating a layer where the hydrophilic heads interact with the water molecules and the hydrophobic tails interact with the air outside the solution, creating space inside (Fig. 8.1). The difference between bubbles and living cells is the location of the water and the orientation of the phospholipids. The membrane of living cells consists of two layers of phospholipids (phospholipid bilayer) but with the hydrophobic tails in contact with each other and the hydrophilic heads forming the surface of the bilayer where it interacts with water on both sides of the membrane (Fig. 8.2). While we focus here solely on the lipid bilayer, it is important to note that cell membranes of living things can be extremely complex and contain many other elements that are critical to maintaining important cellular functions.

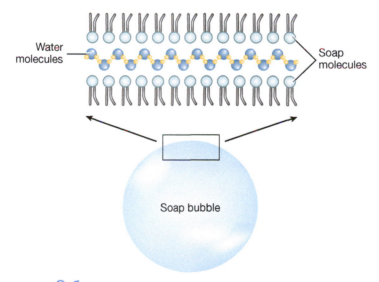

FIGURE **8.1** ● Interaction of soap and water in air forms a bubble.

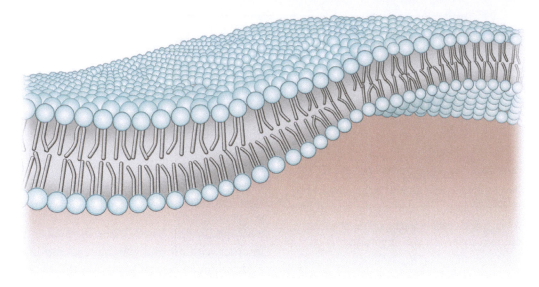

FIGURE **8.2** ● Phospholipid bilayer of a cell membrane.

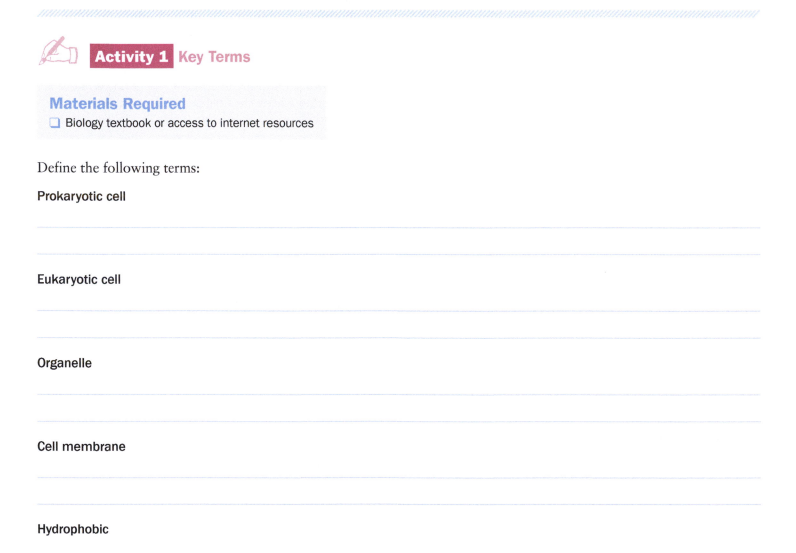

✍ **Activity 1** Key Terms

Materials Required
❑ Biology textbook or access to internet resources

Define the following terms:

Prokaryotic cell

_____

_____

Eukaryotic cell

_____

_____

Organelle

_____

_____

Cell membrane

_____

_____

Hydrophobic

_____

_____

Hydrophilic

_____

_____

Phospholipid bilayer

_____

_____

Mitochondria

_____

Cytoskeleton

_____

Plasma membrane

_____

Nucleus

_____

Rough endoplasmic reticulum

_____

Smooth endoplasmic reticulum

_____

Peroxisome

_____

Ribosome

_____

Cell wall

_____

Chloroplast

_____

Golgi apparatus

_____

Lysosomes

_____

Vacuoles

Centrioles

Microtubules

Activity 2 **Simulation of a Cell Membrane**

Materials Required
- ❏ Flat tray or cookie sheet
- ❏ 10–12 bendable drinking straws
- ❏ Bubble solution:
  - ☐ 4 cups warm water
  - ☐ ½ cup sugar
  - ☐ ½ cup dishwashing soap (Dawn works best)
- ❏ Whisk
- ❏ Thread (about 6 inches total)
- ❏ Paper clips

Note: This experiment will be easier if completed with a partner.

1 To make the bubble solution, use a whisk to mix ½ cup of sugar and 4 cups of warm water (make sure the sugar is completely dissolved). Then add ½ cup of dishwashing soap and use the whisk to mix.

2 Place the tray or cookie sheet on a flat, workable surface. Pour the bubble solution onto the tray or sheet. Whisk until small bubbles form.

3 Using a straw, blow a bubble in the solution.

    a. Why is a bubble a good example of a cell membrane? What do they have in common?

    b. Why is a bubble not a good example of a cell membrane?

4 Holding your thread in both hands, lower it beneath the bubble and draw the thread through the center of the bubble.

    a. Record your observations.

b. How does this represent a possible function of cell membranes?

_____

_____

c. Why is this function essential to cell life?

_____

_____

5 Connect four straws together to form a square by bending one end slightly and inserting it into the end of another until all straws are connected in a square shape, as demonstrated in Figure 8.3.

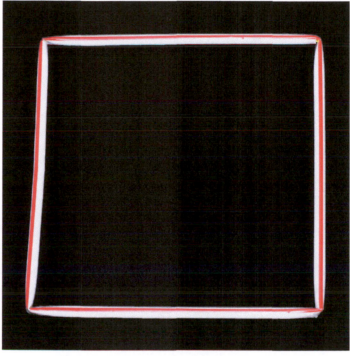

FIGURE **8.3** ● Four straws connected in a square shape.

6 Dip the straws or shape into the bubble solution, and lift to observe the bubble surface.

7 Tie your thread into a circle and lay on the bubble surface inside the straw rectangle. Lift the straws so that the thread is floating on the bubble surface. Place your finger inside of the thread circle to pop the inner bubble surface.

a. What happened to the bubble surface around the thread?

_____

_____

b. Use key terms from this lesson to describe which aspect of cell membranes the thread represents.

_____

_____

c. Use key terms from this lesson to describe how this function is essential to living cells.

_____

_____

8 Remove the string and dip the straw square into the bubble solution to create a new bubble surface within it. Immerse a finger completely in the bubble solution. Insert the finger covered with bubble solution into the surface of the bubble.

  a. What happened?

  _____

  _____

  b. Why did the bubble react that way?

  _____

  _____

  c. Use key terms from this lesson to describe how this is similar to one of the functions of cell membranes and protein receptors.

  _____

  _____

9 Insert a finger not covered in bubble solution into the bubble surface.

  a. What happened?

  _____

  _____

  b. Why did the bubble react that way?

  _____

  _____

10 Blow another bubble in the solution to represent a new cell membrane. Try to insert the paper clip into the bubble (do not dip the paper clip into the bubble solution first).

  a. What happened?

  _____

  _____

  b. Use key terms from this lesson to describe how this mimics the function of a cell membrane.

  _____

  _____

11 Make another hole using the circled thread (as in step 5) and use the hole to insert the paper clip.

  a. What aspect of cell membranes does this represent?

  _____

  _____

  b. Use key terms from Activity 1 to describe how this function is essential to living cells.

  _____

  _____

Activity 3 Chart of Differences

Materials Required
No materials required for this exercise

1 What are the differences between prokaryotic and eukaryotic cells? Complete Table 8.1 to answer this question.

TABLE **8.1** Differences between Prokaryotes and Eukaryotes

	Prokaryotes	Eukaryotes
Number and complexity of organelles		
Extant and complexity of cytoskeleton		
Location of DNA		
Size of the cell		

Activity 4 Drawing Cells

Materials Required
☐ Colored pencils

1 In the space provided, draw and label a plant cell and an animal cell.

2 In your drawing, be sure to label the following: mitochondria, cytoskeleton, plasma membrane, nucleus, rough ER, smooth ER, peroxisome, ribosome, cell wall, chloroplast, Golgi apparatus, lysosomes, vacuoles, and centrioles.

Activity 5 Synthesis

Materials Required
☐ Colored pencils

Now let's tie it all together.

1 How are bubbles similar to cell membranes?

2 How are bubbles different from cell membranes?

_____

_____

3 How do prokaryotes and eukaryotes differ in terms of their compartmentalization?

_____

_____

4 How could differences in compartmentalization affect reaction efficiency?

_____

_____

5 Which structures are found only in plant cells?

_____

6 Which structures are found only in animal cells?

_____

7 Imagine that you are looking through a microscope at plant cells, specifically, epidermal cells. Sketch one of these cells in the space provided. How does this cell differ from the one you drew in Activity 4? (**Hint:** Consider numbers and types of organelles, cell-cell junctions, etc.)

_____

_____

_____

_____

_____

8 Imagine that you are looking through a microscope at some animal cells, specifically heart muscle cells. Sketch one of these cells in the space provided. How does this cell differ from the one you drew in Activity 4? (*Hint:* Consider numbers and types of organelles, cell-cell junctions, etc.)

9 Pretend you are looking through a microscope at cells from different human organs. In which organs would you expect to find cells with the most mitochondria? The most lysosomes? The most ribosomes? Explain your answers.

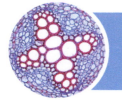

Exercise 9
Cell Structure and Function: Diffusion and Osmosis

Objectives

At the completion of this exercise, students will be able to:

1 Define and apply key terms associated with diffusion and osmosis.

2 Distinguish the fundamental differences between diffusion and osmosis.

3 Describe the effect of temperature and medium on rates of diffusion.

4 Describe how diffusion relates to molecules moving across the cell plasma membrane.

5 Apply concepts of tonicity to osmosis.

(i) Background Information

All matter is comprised of atoms and molecules that are constantly moving, whether as solids, liquids, or gases. As heat is added, the atoms and molecules move faster.

Diffusion is the movement of molecules from an area of higher concentration to an area of lower concentration. The movement is due to molecular collisions, which occur more frequently in areas of higher concentration.

Cells are composed mostly of water but there are many membranous compartments and dissolved molecules that create a gel-like consistency. In Activity 3, you will explore the rate of diffusion in a gelatin.

Unlike animal cells, plant cells are surrounded by a cell wall. The cell wall protects the cell and acts much like a tire, enabling the cell to become rigid, or turgid, when there is pressure inside the cell. When the pressure inside plant cells drops, the cells become flaccid and the plant wilts. This pressure, called turgor pressure, is due to osmosis. In Activity 4, a potato strip will become soft and limp when water leaves the cells, and it will become firm and rigid when water enters.

✎ **Activity 1** Key Terms

Materials Required
☐ Biology textbook or access to internet resources

Define the following terms:

Diffusion

_____

_____

Facilitated diffusion

_____

_____

Osmosis

_____

_____

_____

Active transport

Tonicity

Isotonic

Hypertonic

Hypotonic

Concentration gradient

Activity 2 Rate of Diffusion in Water

Materials Required
- ❑ 2 glasses that are approximately 6.5 cm (approximately 2.5 in.) in diameter and at least 10 cm (4 in.) tall
- ❑ Cold tap water (this can be prepared by cooling water in a refrigerator or by putting ice cubes in water; the ice must be removed before using the water in the experiment that follows)
- ❑ Hot tap water
- ❑ Food coloring

1 Before you begin, read the experiment that follows and create a hypothesis regarding the rate of diffusion of food coloring in hot and cold water.

Null hypothesis:

Alternative hypothesis: _____

_____

Explanation for alternative hypothesis: _____

_____

2 Add approximately 9 cm of cold tap water (3.5 in.) to one glass and 9 cm of hot tap water to another.

3 Place the glasses in a location where they can remain undisturbed. Allow the glasses to remain undisturbed for at least 2 minutes before proceeding to the next step.

4 Carefully place one small drop of food coloring on the surface of the water in the center of each glass. Be careful to not touch the glasses or disturb the water when adding the food coloring. For accurate results the water should be as still as possible.

5 Observe the pattern of diffusion from the side of each glass. Do your observations for the two glasses differ?

6 Estimate the percent of water in the glass that is colored every 30 seconds for 5 minutes. Initially (time = 0 sec.), 0 percent will be colored. Round your answers to the nearest 10 percent (10%, 20%, 30%, etc.). Graph your results in Figure 9.1.

FIGURE **9.1** ● Rate of diffusion in water.

7 Record the amount of time that it takes for the food coloring to reach equilibrium in each glass. Depending on conditions, this may occur in a few minutes or it may take several hours.

8 In which glass does diffusion occur the fastest?

9 Which of your hypotheses is supported by your observations?

10 Use the key terms from this lesson to describe how temperature affects the rate of diffusion.

Activity 3 Rate of Diffusion in Gelatin

Materials Required
- ❏ 1 box of lemon or other light colored gelatin dessert (such as Jell-O)
- ❏ 2 or 3 glasses that are approximately 6.5 cm (approximately 2.5 in.) in diameter and at least 10 cm (4 in.) tall
- ❏ Tap water
- ❏ Food coloring
- ❏ Plastic wrap
- ❏ Rubber band

1 Before you begin, read the experiment that follows and create a hypothesis regarding the rate of diffusion of food coloring in gelatin compared to that in water.

Null hypothesis:

Alternative hypothesis:

Explanation for alternative hypothesis:

2 Prepare a gelatin dessert mix according to the manufacturer's instructions and add approximately 9 cm (3.5 in.) of the liquid to a glass and allow it to harden in a refrigerator.

3 After the gelatin hardens, remove it from the refrigerator and place it in an area where it will not be disturbed for several hours.

4 Add approximately 9 cm of water to a second glass and place it next to the glass containing the gelatin.

5 Both glasses should remain undisturbed for approximately 2 hours so that they are at room temperature when the experiment begins.

6 When the water and gelatin are at room temperature, carefully place one drop of food coloring on the surface of the water and another drop on the surface of the gelatin. The drops should be located in the center of the glasses.

7 To prevent the gelatin from drying, cover the glass containing the gelatin with plastic wrap and secure the plastic wrap with a rubber band. This is necessary because the experiment may take more than one day.

8 Record the amount of time that it takes for the color to diffuse throughout the water.

9 Allow the gelatin to remain undisturbed, checking it at least daily for 3 days, observing the pattern of diffusion. For each observation, record the diameter and depth of the diffused spot and estimate the percentage of gelatin that contained diffused food coloring.

10 In which glass (treatment) does diffusion occur the fastest?

11 Which of your hypotheses is supported by your observations?

12 Use the key terms from this lesson to describe the effect of medium (Jell-O/water) on the rate of diffusion?

13 How does the rate of diffusion in a gelatin at room temperature compare with the rate of diffusion in water at the same temperature?

Materials Required
- ❑ Tap water
- ❑ Table salt
- ❑ Tablespoon (for measuring)
- ❑ Knife (for cutting a potato)
- ❑ 1 potato

1 Before you begin, read the experiment that follows and create a hypothesis regarding what will happen to the firmness of the potato placed in freshwater and the potato placed in salt water.

Null hypothesis: _____

_____

Alternative hypothesis: _____

_____

Explanation for alternative hypothesis: _____

_____

2 Mix one tablespoon (15 mL) of salt with 2 cm of water in a glass. A second glass should contain an equal amount of pure water.

3 Cut two strips of potato about the size of a french fry. They should be no thicker than 0.5 cm.

4 Place one of the strips in the salt water and the other strip in the pure water. Be sure that the water covers the potato strips. Sketch your observations in the circle labeled "Potato before treatment."

5 Leave the strips in the water for 60 minutes. After 60 minutes, examine each strip and record your observations of the firmness of the strips.

6 Sketch the results of the salt treatment and the pure water treatment in the circles labeled "Potato after salt water solution" and "Potato after pure water solution," respectively.

Potato before treatment   Potato after salt water solution   Potato after pure water solution

7 Using the key terms from this exercise, describe the results of both treatments.

_____

_____

_____

_____

Activity 5 Cellular Response to Hypertonic and Hypotonic Solutions

Materials Required
- ❏ 2 chicken eggs
- ❏ Jar (large enough to hold 2 eggs)
- ❏ Vinegar
- ❏ 2 clear drinking glasses
- ❏ Tap water
- ❏ Corn syrup

Note: A chicken egg is a large cell covered by two membranes. The hard outer membrane (the shell) protects the inner semipermeable membrane.

1 Before you begin, read the procedure that follows and develop hypotheses about how you predict the eggs will respond to the different treatments.

Null hypothesis: _____

_____

Alternative hypothesis: _____

_____

Explanation for alternative hypothesis: _____

_____

_____

2 To conduct this experiment the shell must be removed. To remove the shell place the two eggs in a jar of vinegar large enough to cover the two eggs with the liquid. Allow to stand for three days. Because the inner semipermeable membrane is fragile, take care not to puncture or tear the membrane when moving the eggs.

3 Place each egg in a clear glass. Cover the first egg in water. Cover the second egg in corn syrup. Allow the eggs to sit in the solutions for 8 hours.

4 Which solution is hypertonic and which is hypotonic? Explain.

_____

_____

_____

5 Measure each egg and record the results.

6 Explain the results.

7 Which egg shows the greatest change? Why?

8 Now reverse the process. Place the egg that was originally in the water into a glass with corn syrup and place the egg that was in the corn syrup into a glass with water. Let stand for 8 hours.

9 Measure each egg and record the results.

10 Using key terms from this lesson, describe your results.

11 Which egg shows the greatest change?

12 Use key terms from Activity 1 to explain the difference between the responses of the two eggs.

Activity 6 Synthesis

Materials Required
No materials required for this exercise

Now let's tie it all together.

FIGURE **9.2** • Red blood cell in solutions of various tonicity.

1 Students frequently conflate the tonicity of the solution a cell is placed in with the contents of the cell—don't make that mistake! Figure 9.2 depicts a red blood cell placed in three different solutions. Distinguish between the tonicity of the cell, and the tonicity of each solution. State the tonicity of each solution under each image.

2 Match each activity below with the correct description of the net movement of molecules across a cell membrane. The direction of net movement choices (A or B) may be used more than once for each answer.

A = movement from high to low concentration
B = movement from low to high concentration

_____ 1. Simple, passive diffusion

_____ 2. Facilitated diffusion

_____ 3. Active transport

_____ 4. Water with regards to the water concentration gradient during osmosis

_____ 5. Water with regard to the solute concentration gradient during osmosis

3 In osmosis, water always moves toward the _____ solution; that is, it moves toward the solution with the

_____ solute concentration.

a. isotonic, greater

b. hypertonic, greater

c. hypertonic, lesser

d. hypotonic, greater

e. hypotonic, lesser

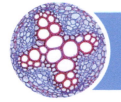

Exercise 10
Energy Conversion in Eukaryotic Cells: Respiration

Objectives

At the completion of this exercise, students will be able to:

1 Define and apply key terms associated with cellular respiration.

2 Describe basic ideas and components of cellular respiration.

3 Compare and contrast aerobic and anaerobic respiration.

4 Integrate basic ideas about respiration through observing the fermentation process in yeast.

ⓘ Background Information

In the same way that you can't drop a banana into the gas tank of a car and expect the car to go anywhere, your body's cells can't directly burn the components of that banana you just ate in order to get stuff done. The energy that cells need to do work does come from the banana you just ate, but before the cells can do anything with it, the banana must first be broken down by your digestive system into different types of sugars, amino acids, and fatty acids. Then, in order to be used by the cell, these must be converted into a usable form of energy (ATP). Cellular respiration is the process cells use to break down complex molecules into simple sugars and convert them to ATP. These chemical reactions are considered catabolic, because they break large molecules into smaller ones, releasing energy in the process.

✏️ **Activity 1** Key Terms

Materials Required
☐ Biology textbook or access to internet resources

Define the following terms:

Aerobic respiration

_____

_____

Fermentation

_____

_____

ATP

_____

_____

_____

Glycolysis

_____

_____

Reduction reaction

_____

_____

Oxidation reaction

_____

_____

Electron acceptor

_____

_____

Electron transport chain

_____

_____

Krebs/citric acid cycle

_____

_____

Catabolic reaction

_____

_____

Anabolic reaction

_____

_____

Activity 2 Reduction/Oxidation (Redox) Reactions

Materials Required
No materials required for this exercise

The chemical reaction that takes place when paper burns (combustion) is a redox reaction. As the paper burns, chemical energy stored in the paper (a complex sugar) reacts with an oxidizing agent (oxygen) to produce carbon dioxide, water, heat, and light (Fig. 10.1).

$$C_6H_{12}O_6 \;+\; 6\,O_2 \;\rightarrow\; 6\,CO_2 \;+\; 6\,H_2O \;+\; ATP$$

| glucose | oxygen gas | carbon dioxide gas | water | energy |

FIGURE **10.1** ● Redox reaction.

As with combustion, the reactions involved in cellular respiration are catabolic. They break the high-energy bonds in large molecules into smaller ones, releasing energy in the process (Fig. 10.2).

$$C_6H_{12}O_6 \;+\; 6\,O_2 \;\rightarrow\; 6\,CO_2 \;+\; 6\,H_2O \;+\; light \;+\; heat$$

| glucose | oxygen gas | carbon dioxide gas | water | energy |

FIGURE **10.2** ● Cellular respiration.

The oxidizing agent (electron acceptor) in both reactions is molecular oxygen (O_2). Molecules that lose an electron are oxidized. These molecules also lose a proton. Oxidized compounds have many C-O bonds.

As illustrated by the difference between combustion and cell respiration, many small controlled reactions allow the cell to use the energy from breaking the bonds of each glucose molecule in order to capture as many electrons and create as many ATP molecules as possible. The first step in cell respiration is glycolysis—the breakdown of your six-carbon sugar molecule. The energy released from breaking the carbon-carbon bonds in this molecule releases electrons that can be stored (Figs. 10.3 and 10.4).

> **Note:** Molecules that gain an electron are reduced. These molecules also gain a proton (H^+). Reduced molecules often have many C-H bonds.

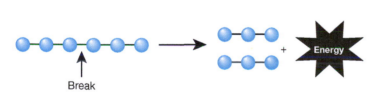

FIGURE **10.3** ● Energy released from breaking carbon-carbon bonds.

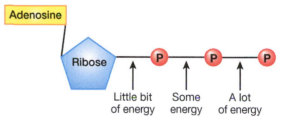

FIGURE **10.4** ● Structure of ATP and the energy contained in its bonds.

As glucose breaks into pyruvate, the electrons that are released enter the electron transport chain, and are used to build the proton gradient that powers ATP synthase, which in turn produces ATP. In the presence of oxygen, pyruvate moves through the Krebs cycle, where more carbon bonds are broken (resulting in the production of more ATP) and then to the electron transport chain, where large amounts of ATP are synthesized (aerobic respiration). However, if there is no electron acceptor (oxygen) present, pyruvate is broken down through fermentation. This is anaerobic respiration. In humans, lactic acid fermentation takes place, whereas in yeast, alcohol formation occurs.

Living cells obtain energy by transferring electrons from an electron donor to an electron acceptor. As illustrated in Figure 10.5, the most efficient way to make ATP is aerobically, through the electron transport chain. During this process, the electron acceptor is reduced and the electron donor is oxidized.

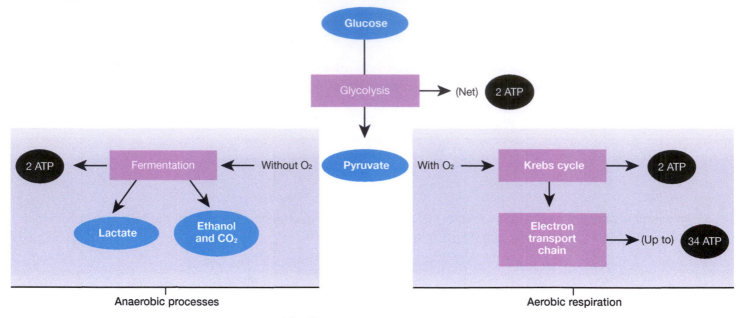

FIGURE **10.5** ● Anaerobic versus aerobic respiration.

1 Using key terms in this lesson, list the ways in which combustion and cell respiration are similar.

_____

_____

_____

_____

2 In Figure 10.2, identify which reactants become oxidized and which become reduced.

_____

_____

_____

_____

Activity 3 Observing Fermentation

Materials Required

☐ 1 balloon
☐ Small funnel (plastic or made with paper will work fine)
☐ 1 tbsp. active yeast
☐ 1 tsp. sugar
☐ ¼ cup warm water (feels hot but not too hot to place finger in)
☐ Ruler

Note: Do not inflate balloon prior to starting experiment.

1 Combine yeast and sugar together and set aside.

2 Using the funnel, add the yeast/sugar mixture to the balloon, and then add water. Tie the balloon shut. Lay the balloon flat on work surface.

3 Measure the length of the balloon with a ruler.

Length: _____

4 Measure the width of the balloon with a ruler.

Width: _____

5 Find a warm place. Set the balloon there, and let it sit for 45 minutes.

6 Remove the balloon, and measure its length and width again.

Length: _____

Width: _____

7 Describe what changes took place, if any.

_____

_____

8 How do you explain the difference in size?

_____

_____

9 What are the reactants and products of fermentation? What are the reactants and products in this experiment?

_____

_____

_____

_____

Materials Required

No materials required for this exercise

Remember all the energy stored in those C-H chemical bonds? When the electrons in those bonds transfer from electron donors to electron acceptors the mitochondria in your cells can harness that energy. This series of redox reactions is called the electron transport chain, because it moves electrons from an electron donor (NADH) to an electron acceptor (O_2) in a stepwise fashion. Ultimately, these reactions create a surplus of protons (H^+) on one side of the mitochondrial inner membrane. In the same way that water flows through a turbine at the base of a hydroelectric dam, all those protons go from a higher concentration on one side of the membrane to a lower concentration on the other side of the membrane, crossing through a membrane-bound protein (ATP synthase) that spins at 50 revolutions per second! The energy generated by this spinning action binds a free phosphate (Pi) to adenosine diphosphate (ADP) to make adenosine triphosphate (ATP). This process is called oxidative phosphorylation.

1 Using your textbook and internet resources, illustrate Figure 10.6, which shows the electron transport chain. Be sure to include the following in your diagram: Complex I, Complex II, Complex III, Complex IV, mitochondrial matrix, cristae membrane, and the intermembrane space. Identify which components comprise the electron transport chain, and which are involved in oxidative phosphorylation.

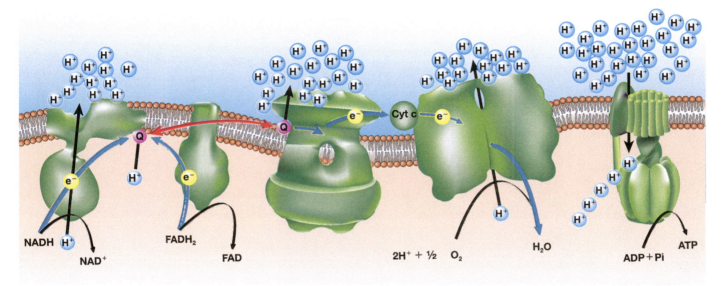

FIGURE **10.6** ● Electron transport chain.

Activity 5 Synthesis

Materials Required
No materials required for this exercise

Now let's tie it all together.

1 Where in the cell does glycolysis take place?

_____

_____

2 In glycolysis, which molecules are the reactants (*Hint:* What goes in?)

_____

_____

3 What are the products of glycolysis?

_____

_____

4 Which element must be present for aerobic respiration to take place?

_____

_____

5 Where in the cell does the Krebs/citric acid cycle take place?

_____

_____

6 What are the reactants in the Krebs/citric acid cycle? (*Hint:* What goes in?)

_____

_____

7 What are the products of the Krebs/citric acid cycle? (*Hint:* What comes out?)

_____

_____

8 What are the reactants and products of the electron transport chain?

_____

_____

9 What are the reactants and products of oxidative phosphorylation?

_____

_____

10 Cyanide (CN^-) poisoning stops cellular respiration by inhibiting the action of cytochrome c oxidase enzymes in the mitochondria. Specifically, cyanide blocks complex IV of the electron transport chain. What does this suggest about oxygen's role as an electron acceptor?

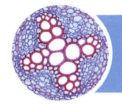

Exercise 11
Energy Conversion in Eukaryotic Cells: Photosynthesis

Objectives

At the completion of this exercise, students will be able to:

1 Define and apply key terms associated with photosynthesis.
2 Diagram the overall process of photosynthesis.
3 Describe the role of photons in photosynthesis.
4 Distinguish between autotrophic and heterotrophic organisms.
5 Write a summary equation for photosynthesis.
6 Compare and contrast photosynthesis and cellular respiration.
7 Illustrate light-dependent, and light-independent reactions.
8 Describe carbon fixation.
9 Explain why rubisco is the most important enzyme on earth.

ⓘ Background Information

Photosynthesis is the process by which autotrophs (plants, algae, and some protists and prokaryotes) use sunlight to convert carbon dioxide (CO_2) and water (H_2O) into sugar ($C_6H_{12}O_6$), which can then be converted to ATP to power cellular activities.

> **Note:** Photosystem II comes before Photosystem I in the process of photosynthesis. Photosystem I was simply discovered first.

Activity 1 Key Terms

Materials Required
☐ Biology textbook or access to internet resources

Define the following terms:

Chlorophyll

_____

_____

Stoma

_____

_____

ATP

_____

_____

NADPH

_____

_____

Chloroplast

Thylakoid

Sugar

Starch

Carbohydrate

Photon

Carbon fixation

Photosystem II

Photosystem I

Z scheme

Rubisco

Calvin cycle

Light reaction (light-dependent reaction)

Dark reaction (light-independent reaction)

Electron transport chain

Autotroph

Heterotroph

Oxidative phosphorylation (mitochondria)

Photophosphorylation (chloroplasts)

Materials Required

- ❏ Houseplant
- ❏ Small saucepan
- ❏ Medium-sized saucepan
- ❏ Kitchen stove or burner
- ❏ 2 small, clear juice glasses (made of glass—must be able to withstand very hot water)
- ❏ Alcohol (isopropyl; from the drug store)
- ❏ Iodine (from the drug store; commonly found as wipes in first aid kits)
- ❏ Aluminum foil
- ❏ Kitchen tongs, tweezers, or fork (used to move leaves between solutions without touching them)
- ❏ Colored pencils

> **Note:** Activity 2 requires wrapping some of the leaves of a houseplant in foil for 3–7 days before conducting the experiment.

1 Carefully wrap a few leaves of the houseplant with aluminum foil to block access to sunlight (or artificial light). Allow the other leaves on the same plant normal access to light. Take care not to damage the leaf as you wrap it in foil. Let the plant remain in this condition for 3 to 7 days.

2 Develop hypotheses about how much sugar is created by the leaf kept in the dark versus the leaf kept in the light.

Null hypothesis: _____

_____

_____

_____

Alternative hypothesis: _____

_____

_____

_____

3 Remove a leaf that was exposed to the light. If it is large, you may have to cut it so that it can fit flat on the bottom of the juice glass.

4 Add 300 mL (~1⅓ cups) of water to the small saucepan.

5 Place the saucepan with water on the stove and bring it to a boil.

6 Place the leaf in the boiling water for 1 minute. This denatures and inactivates the chlorophyll.

7 Remove the saucepan from the burner.

8 Add 75 mL (~⅓ cup) of alcohol to one of the juice glasses. Very carefully place the leaf in the alcohol (in the juice glass), and then place the glass in the saucepan with the very hot (nearly boiling) water. This is called a water bath. This step dissolves away the chlorophyll.

9 Incubate in the water bath for 5 minutes.

10 Very carefully remove the leaf from the alcohol and place it in the bottom of the second juice glass. Cover the leaf with water and 5 to 10 drops of iodine, or enough to cover most of the surface of the leaf. The iodine stains the sugar a dark blue or black color, but remains yellow where there is no sugar.

Note: If you are using iodine from wipes, you may need to squeeze the drops out of the gauze.

11 Record your observations and drawings in the spaces provided.

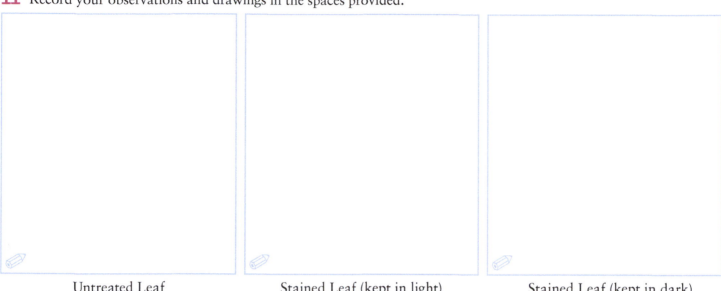

Untreated Leaf   Stained Leaf (kept in light)   Stained Leaf (kept in dark)

12 Repeat steps 4 through 10 with a leaf that was covered in aluminum foil.

13 How does the pattern of chlorophyll differ between the leaf kept in the dark versus the leaf kept in the light?

14 How does the pattern of sugar formation relate to the amount of chlorophyll?

15 Using a single complete sentence, write a conclusion about the relationship between chlorophyll and photosynthesis. What evidence supports your conclusion?

16 Using a single complete sentence, write a conclusion about the relationship between exposure to light and photosynthesis. What evidence supports your conclusion?

17 How did you control for the differences between the treatments in your experiment?

_____

_____

_____

_____

_____

 Activity 3 **Understanding Photosynthesis**

Materials Required
❑ Colored pencils
❑ Biology textbook or access to the internet

1 In the space to the right of Figure 11.1, make your own rendition of the overall process of photosynthesis. If recommended by your instructor, use other sources to modify your figure.

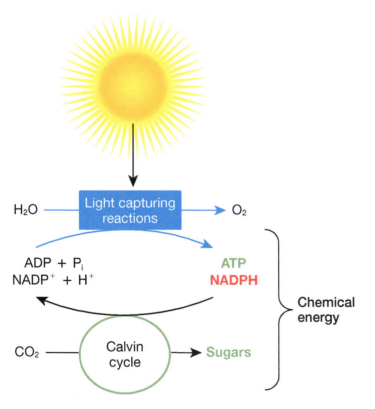

FIGURE **11.1** ● Process of photosynthesis.

$$\text{Light} + 6 O_2 + H_2O \rightarrow (CH_2O)_n + O_2$$

energy   carbon   water   sugar   oxygen
dioxide

FIGURE **11.2** • Summary of the reactants and products of photosynthesis.

The first part of the equation involves harvesting light (Fig. 11.2). In this process, energy in the form of light is captured and temporarily stored in ATP and NADPH. The stored energy in ATP and NADPH is then used to make sugar from the CO_2 in a process called carbon fixation.

The light harvesting part of the equation starts when light energy is absorbed by chlorophyll, a pigment that is embedded in the chloroplast membranes of plant cells. The energy absorbed by the photosystem excites the electron in a chlorophyll molecule and moves it to a higher energy state, allowing the electrons to be more easily transferred to electron acceptors ($NADP^+$). To replace the electrons that are lost by the chlorophyll molecules, water (H_2O) is split, yielding electrons, protons (H^+), and diatomic oxygen gas (O_2). The energy created by the electron acceptors is used to pump hydrogen ions (protons) across the thylakoid membrane. This proton gradient is used to make ATP from ADP and P_i (inorganic phosphate). The $NADP^+$ is turned into NADPH with the addition of one proton (H^+) and two electrons.

2 In the space below Figure 11.3, make your own rendition of the light reaction part of photosynthesis. If recommended by your instructor, use other sources to modify your figure.

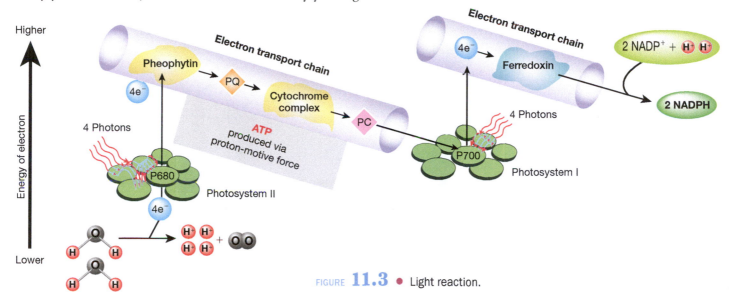

FIGURE **11.3** • Light reaction.

Carbon fixation uses the electrons and energy from the ATP and NADPH created by the light reaction to make carbon-carbon bonds in the dark reaction part of photosynthesis (Fig. 11.4). Think of this as a three-part process:

1. CO_2 comes in.
2. Energy comes in.
3. Electrons come in.

The reaction starts with a five-carbon molecule, and rubisco, the most abundant enzyme on planet Earth, catalyzes the conversion of CO_2 and the five-carbon sugar (ribulose bisphosphate) to make two c-carbon molecules, (3-phosphoglycerate, or 3PGA). Then each of the 3PGA molecules are reduced (electrons are added to them) to make three-carbon sugars (glucose or fructose). But to reconstitute the five-carbon molecule regenerate the rubisco (RuBP), ATP is added to some of the remaining G3P so the process can start again.

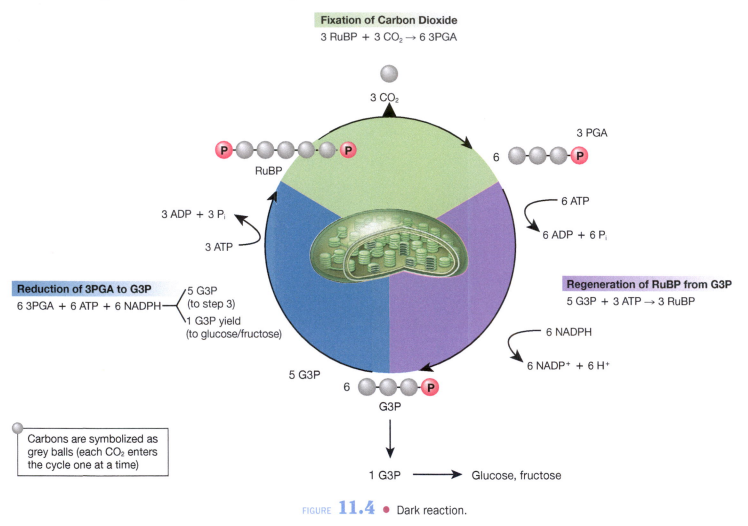

Fixation of Carbon Dioxide
$3\ RuBP + 3\ CO_2 \rightarrow 6\ 3PGA$

3 CO_2

3 PGA

RuBP

6 ATP

3 ADP + 3 P_i

6 ADP + 6 P_i

3 ATP

Reduction of 3PGA to G3P

$6\ 3PGA + 6\ ATP + 6\ NADPH$

5 G3P (to step 3)

1 G3P yield (to glucose/fructose)

Regeneration of RuBP from G3P

$5\ G3P + 3\ ATP \rightarrow 3\ RuBP$

6 NADPH

6 NADP⁺ + 6 H⁺

5 G3P

G3P

Carbons are symbolized as grey balls (each CO_2 enters the cycle one at a time)

1 G3P ⟶ Glucose, fructose

FIGURE **11.4** • Dark reaction.

3 Using Figure 11.4 as a reference, make your own rendition of the dark reaction part of photosynthesis in the space below. If recommended by your instructor, use other sources to modify your figure.

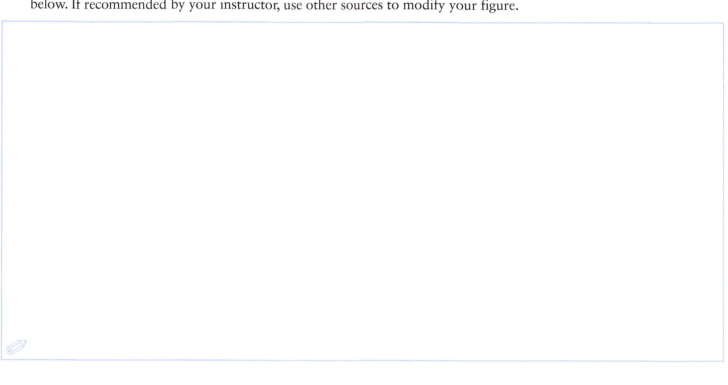

 **Activity 4** Teach It!

Materials Required
- ❏ Colored pencils, pens, or markers
- ❏ Poster board
- ❏ Biology textbook or access to the internet

1 Using images and materials from your textbook or the internet as your guide, diagram the process of photosynthesis on a poster board.

2 On the poster and in your discussions be sure to cover the following topics:

What overall reaction occurs?

What gets oxidized?

What gets reduced?

Where do the light-dependent reactions occur?

Where do the light-independent reactions occur?

Where do the electrons for photosystem II come from?

Where do the electrons for photosystem I come from?

What are the phases of the Calvin cycle?

What goes in and what comes out of each phase?

Why is rubisco (and the reaction it catalyzes) so important?

3 Teach five different people about respiration and have them initial your poster.

 **Activity 5** Synthesis

Materials Required
No materials required for this exercise

Now let's tie it all together.

1 What is the purpose of the light cycle?

_____

_____

2 What is the purpose of the dark cycle?

_____

_____

3 Why are plants green?

_____

_____

4 When you were younger you may have heard that "animals breathe in oxygen and exhale carbon dioxide, while plants take in carbon dioxide and exhale oxygen." In what ways is this statement true? In what ways is it false?

_____

_____

_____

5 Describe how respiration can be characterized as photosynthesis in reverse, and vice versa.

_____

_____

_____

6 Using Figure 11.5, compare and contrast respiration and photosynthesis and discuss how they complement each other.

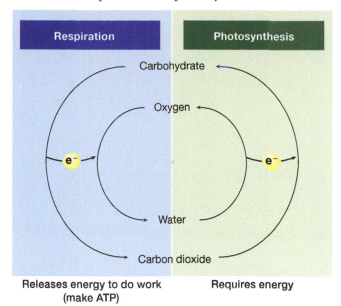

FIGURE **11.5** ● Comparison of respiration and photosynthesis.

Genetics: Heredity

· · · · · · · · · ·

Exercise 12
Mendelian Genetics: Monohybrid and Dihybrid Crosses

Objectives

At the completion of this exercise, students will be able to:

1 Define and apply key word stems of genetics terminology.

2 Define and apply fundamental terms of modern genetics.

3 Define *monohybrid cross* and illustrate an example.

4 Explain how the law of segregation describes the inheritance of a single trait.

5 Define *dihybrid cross* and illustrate an example.

6 Describe how the law of independent assortment applies to a dihybrid cross.

7 Explain how a test cross can be used to determine an organism's genotype.

ⓘ Background Information

Understanding the word stems of key genetics terms is helpful to understanding important processes of modern genetics. Table 12.1 provides some examples of prefixes used in genetics.

Monohybrid crosses depict a mating between two individuals who have different alleles at a single locus: one gene (locus), two alleles per locus (each parent has two alleles for a single gene). Dominant alleles are commonly designated by capital letters; recessive alleles are lower case. The Punnett square is a simple way to determine and visualize the possible genotypes of the offspring of two parents of known genotype. It reveals all the possible combinations of alleles in a genetic cross. Genotypes can be used to predict phenotypes.

A dihybrid cross is a 2-locus cross, with two alleles per locus. To calculate a dihybrid cross, your Punnett square must have 16 squares because there are four possible allele pairs that can be matched.

TABLE **12.1** Prefixes Used in Genetics

Prefix	Meaning
co-	together
mono-	one
di-	two
poly-	many
pleio-	more
homo-	same
hetero-	different

Activity 1 Key Terms

Materials Required
❑ Biology textbook or access to internet resources

Define the following terms:

Genotype

_____

_____

Phenotype

_____

_____

Locus

Allele

Gamete

Diploid

Haploid

Dominant allele

Recessive allele

Codominant

Punnett square

Genetic linkage

Homozygote

Heterozygote

Monohybrid cross

_____

_____

Dihybrid cross

_____

_____

Law of Dominance

_____

_____

Law of Segregation

_____

_____

Law of Independent Assortment

_____

_____

Test cross

_____

_____

Activity 2 **Monohybrid Crosses**

Materials Required
No materials required for this activity

Scenario A

In goats, black fur is dominant to red fur. The gene for this trait can be described as Bb (if you forgot why this is so, go back and reread the background information at the beginning of this exercise). A homozygous black female goat (nanny) is crossed with a heterozygous male (billy).

1 What is the genotype and phenotype of the nanny?

_____

_____

2 What is the genotype and phenotype of the billy?

_____

_____

3 Using this information, fill in the Punnett square A to predict the potential genotypes of offspring that could result from this cross.

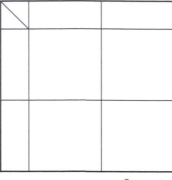

Punnett square **A**

4 What is the probability of this cross producing a black kid (baby goat) that is homozygous for this trait? Explain your answer.

_____

_____

_____

5 What is the probability of the goats having a red kid, i.e., what is the proportion of red kids?

_____

6 What is the probability of the goats having a black kid?

_____

Scenario B

A female (Dd) black dog has three black puppies and one white puppy.

1 What are the possible genotypes of the father of the litter?

_____

2 Use Punnett square B to draw the cross between one of the father genotypes and the mother.

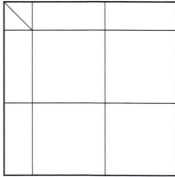

Punnett square **B**

3 What are the possible genotypes for the puppies in the litter?

_____

_____

4 How can there be more than one possible genotype for an observed phenotype?

_____

_____

5 Describe how you could design a test cross to positively identify the genotype of the father.

_____

_____

_____

_____

_____

_____

Scenario C

Assume color in salamanders follows simple dominance and recessive patterns. A green salamander of unknown genotype pairs with a white homozygous recessive salamander (gg).

1 What are the two possible genotypes of the green salamander?

_____

_____

2 Complete Punnett squares C and D for the two possible genotypes.

Punnett square **C**   Punnett square **D**

Activity 3 Dihybrid Crosses

Materials Required
☐ Colored pencils

A heterozygous yellow bird (Yy) has a homozygous short bill (bb) and it pairs with a homozygous white bird (yy) with a heterozygous long bill (Bb). The Punnett square to calculate possible genotypes for this pair would look like this:

	YB	Yb	yB	yb
yB	YyBB	YyBb	yyBB	yyBb
yb	YyBb	Yybb	yyBb	yybb
yB	YyBB	YyBb	yyBB	yyBb
yb	YyBb	Yybb	yyBb	yybb

1 Using the example above, what are all of the possible genotypes for the parents of the heterozygous yellow bird with the homozygous short bill?

_____

_____

2 Using colored pencils or markers, illustrate each of the phenotypes that are predicted by their parental gametes.

	YB	Yb	yB	yb
yB				
yb				
yB				
yb				

Scenario A

A homozygous grey (GG) elephant has heterozygous blue eyes (Bb) and pairs with a homozygous black (gg) elephant with homozygous green (bb) eyes.

1 Draw the dihybrid cross for the pair in Punnett square E.

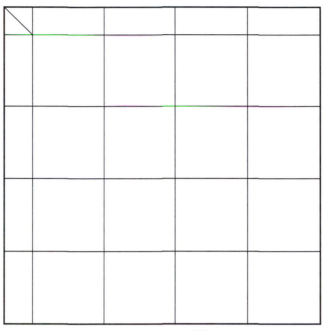

Punnett square **E**

2 What is the probability of the pair producing a grey elephant with blue eyes?

_____

3 What is the probability of a grey elephant with homozygous green eyes?

_____

Now let's tie it all together.

To answer the synthesis questions, you will use a few tosses of a coin to randomly select genotypes of two gargoyle gametes. Then, using these genotypes (gametes), you will create the genotypes and phenotypes that result from the resulting cross (mating). The first coin controls the genotype for wing shape. Heads is dominant (a pointed wing) and tails is recessive (a smooth wing). The second coin controls the genotype for horns. Heads is dominant (curled horns) and tails is recessive (straight horns).

1 Flip each coin twice and record the genotype for the first parent.

_____

2 Flip the coins twice again and record the genotype for the second parent.

_____

3 Draw the cross of your parental gargoyles in Punnett square F.

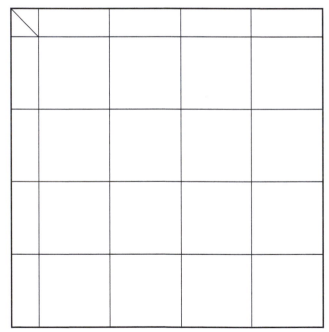

Punnett square **F**

4 Direct your web browser to: **https://www.random.org**. In the random number generator box enter "1" as the minimum value and "16" as the maximum value and click the "generate" button. Record the number here.

_____

5 Randomly choose an offspring from Punnett square F. To do this, circle the number you came up with in question 4 in Punnett square G and write down the genotypes that correspond to the same cell in Punnett square F.

For example, if the random number you generated is "11," you would circle the number 11 in Punnett square G and write down the genotypes from Punnett square F that correspond to position 11 circled in the Punnett square G.

	2	3	4
5	6	7	8
9	10	11	12
13	14	15	16

Punnett square **G**

6 Using the information in your Punnett square (G), what is the probability of getting a genotype that matches your randomly chosen offspring?

7 What is the probability of getting the opposite of that offspring's phenotype?

8 Draw your gargoyle in the space provided.

9 Does Mendel's law of segregation hold true if parental lines differ with respect to two traits? Explain.

_____

_____

10 How are Mendel's laws of independent assortment, segregation, and dominance essential to genetic diversity?

_____

_____

Exercise 13
Mendelian Genetics: Genotypes and the Human Face Exercise

Objectives:

At the completion of this exercise, students will be able to:

1 Define and apply key terms associated with Mendelian genetics.

2 Compare and contrast meiosis and mitosis.

3 Distinguish between genotype and phenotype.

4 Describe crossing-over, independent assortment, and gametogenesis.

5 Explain the physical basis of inheritance.

6 Describe how pedigree charts can be used to determine mode of inheritance.

ⓘ Background Information

The ultimate goal of the Human Genome Project (HGP) was to generate a representative DNA sequence for the human genome's 3 billion base pairs and to identify all human genes. The HGP officially began in 1990, and the last chromosome to be sequenced was completed in May 2006. Now the call is for scientists to: identify genes; develop new tools to allow discovery of the hereditary contributions to common diseases such as diabetes, heart disease, and mental illness; discover new methods for the early detection of disease; develop new technologies that can cheaply sequence the entire genome of any person; and improve the understanding of biological pathways to accelerate drug discovery.

While there are many anticipated benefits of genetic research there are always concerns, such as ethical, legal, and social issues that must be addressed surrounding the availability of genetic data and capabilities. These issues are important to each of us and so we all must become somewhat versed in Genomics 101, so to speak, in order to make wise decisions concerning the use of this powerful information.

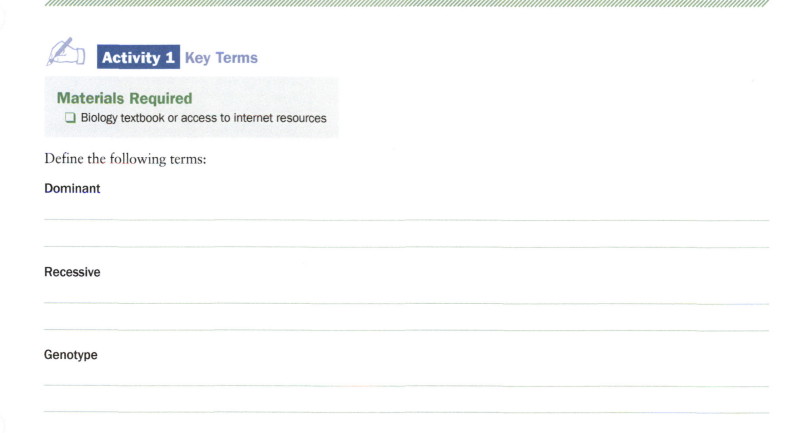

✎ **Activity 1** Key Terms

Materials Required
❑ Biology textbook or access to internet resources

Define the following terms:

Dominant

_____

_____

Recessive

_____

_____

Genotype

_____

_____

Phenotype

_____

_____

Homozygous

_____

_____

Heterozygous

_____

_____

Homologous

_____

_____

Gene

_____

_____

Alleles

_____

_____

Law of independent assortment

_____

_____

Law of segregation

_____

_____

Activity 2 Building a Model Face

Materials Required
❑ Genome template and traits template (included)
❑ Scissors

A genome template (Fig. 13.1) and a traits template (Fig. 13.2) are included in this assignment. They contain chromosomes, genes (loci), and example traits (alleles) needed to build a model face. You will need to cut out all the pieces for each activity before you begin. Part of this assignment will be completed during your model face construction. Some questions you will complete after construction. Refer to Table 13.1 as you work. This table contains the "genetic blueprint" of your model faces and will help you answer the questions.

Note: ALL phenotypes are the result of gene-by-environment interactions and this exercise is a simplification of the actual genetic and environmental mechanisms that eventuate in a person's phenotype. Also note that traits such as eye color, hair color, tongue rolling, widow's peak, detached/attached earlobes, hitchhiker's thumb, etc., once thought to be single-locus, two-allele traits, are likely due to more than one gene or are strongly influenced by environment. While these traits are now known to be polygenic, there are still numerous traits in humans for which single-locus, two-allele systems are very real, including albinism; brachydactyly (shortness of fingers and toes); certain types of breast, ovarian, and colorectal cancers; Huntington's disease; lactase persistence; retinoblastoma; sickle-cell anemia; and many others. For the purpose of this exercise, we're pretending that the traits we are working with are located on only two autosomal chromosomes and the sex chromosomes, and that they are not influenced by environmental variation. See also *Online Mendelian Inheritance in Man* at **http://www.ncbi.nlm.nih.gov/omim** for many more examples.

TABLE **13.1** Genome of Your Model Face

Genotype	Phenotype	Chromosome
Hairline		1
FF	Widow's peak	
Ff	Widow's peak	
ff	Straight hairline	
Nose		2
HH	Broad nose	
Hh	Intermediate width nose	
hh	Narrow nose	
Eyes		1
EE	Brown eyes	
Ee	Brown eyes	
ee	Blue eyes	
Eyebrows		X
N_	Thick	
n	Thin	
Mouth		2
M_R_	Thick red lips	
mmrr	Thin pink lips	
M_rr	Thick pink lips	
mmR_	Thin red lips	
Earlobes		2
AA	Unattached earlobe	
Aa	Unattached earlobe	
aa	Attached earlobe	
Sex		
XY	Male	
XX	Female	

Note: The designation "_" as a genotype means that it doesn't matter which form of the allele is present. For example, the genotype M_R_ means that no matter what the second allele for lip color is (R or r), the person's lips will be red.

1 From the genome template (Fig. 13.1), cut out and gather all the chromosomes. Place each allele on their appropriate locus on the chromosomes according to the template.

2 Organize the chromosomes in pairs: Chromosome 1, Chromosome 2, an X chromosome, and a Y chromosome.

3 Using Table 13.1 and all three sets of chromosomes (Chromosome 1, Chromosome 2, and the sex chromosomes), complete Table 13.2.

TABLE **13.2** Model Facial Traits

	Genotype	Predicted Trait	Principle of Inheritance
Hairline	Ff		
Nose		Intermediate width nose	
Eyes			Complete dominance
Eyebrows			
Mouth			
Earlobes			
Sex			

4 Cut out one of the faces and all of the traits in Figure 13.2.

5 Using your completed Table 13.2, build your model face.

Enhancement Exercises for Biology • **UNIT 4:** Genetics: Heredity

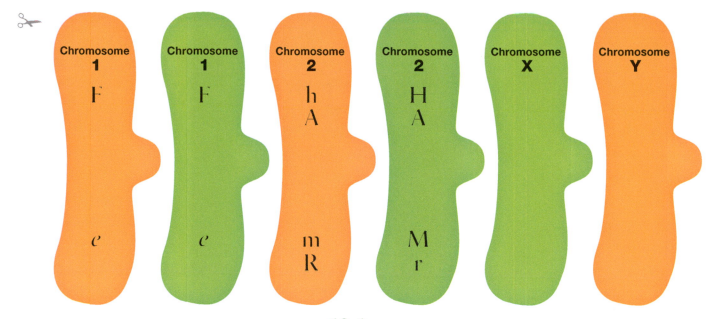

FIGURE **13.1** • Genome template.

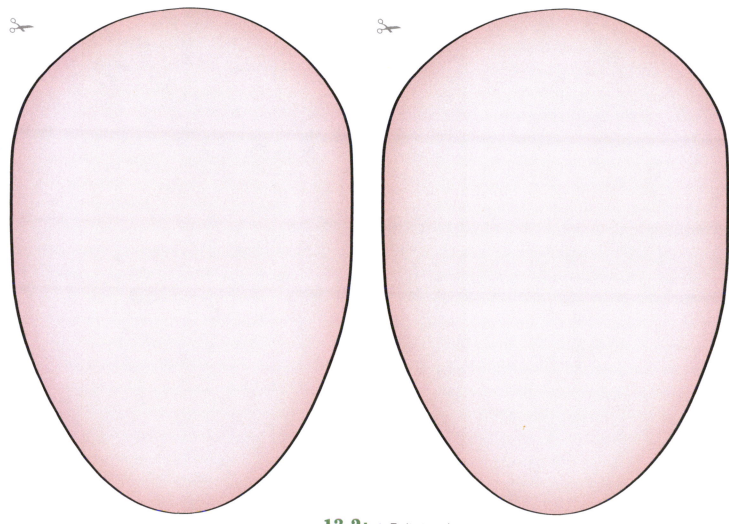

FIGURE **13.2A** • Traits template.

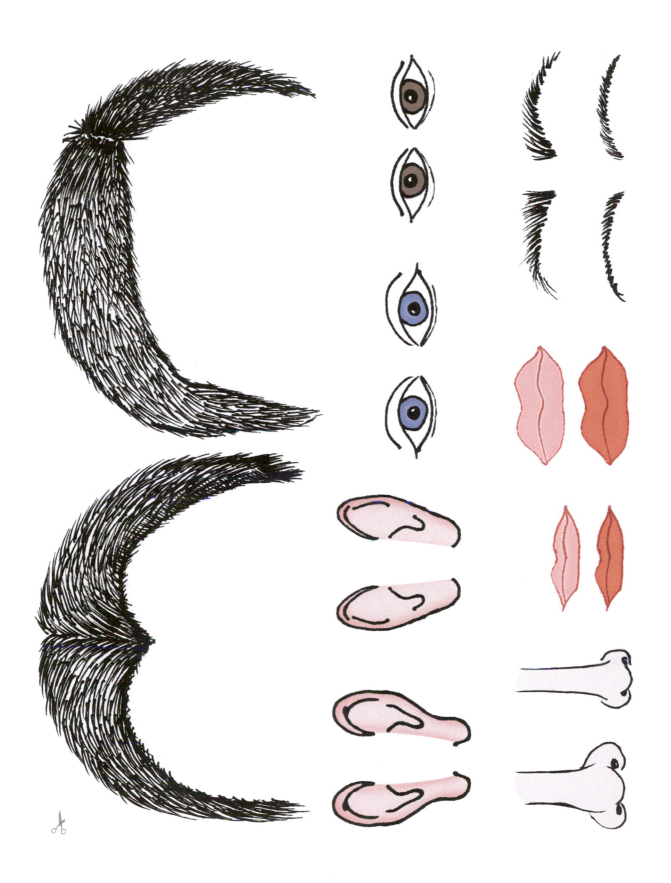

FIGURE **13.2B** • Genetic traits.

Activity 3 Replication

Materials Required
☐ Genome template and traits template (included)

1 Using the chromosomes from Step 2 in Activity 2 and the cutouts from Figure 13.1, replicate your chromosomes in preparation for meiosis.

2 Draw the products of DNA replication in the appropriate spaces in Figure 13.3.

3 Draw your own sex chromosomes including the alleles in the space provided. If you are a female, you'll have two X chromosomes. If you are a male you will have one X and one Y chromosome. Add the alleles for eyebrows to your chromosomes.

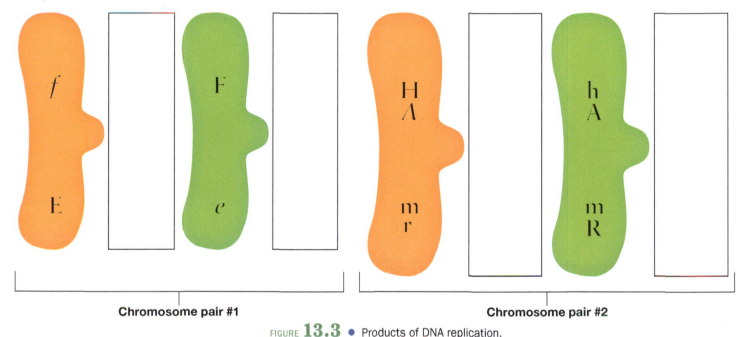

Chromosome pair #1   Chromosome pair #2

FIGURE **13.3** ● Products of DNA replication.

Activity 4 Meiosis

Materials Required
No materials required for this activity

1 Now, go through the process of meiosis using your replicated chromosomes. You'll want to think carefully about how you'll introduce genetic variation (crossing over and independent assortment) and the reduction of chromosome number (diploid is reduced to haploid). As you go through the steps of meiosis with your replicated chromosomes, think carefully about what role genetic variation and reduction of chromosome number plays in the creation of gametes leading to a new generation.

2 In Activity 2, you carried out meiosis. Illustrate how crossing-over events, independent assortment, and chromosome reduction could have taken place in Step 2 of Activity 3 in the appropriate gametes (for this exercise, only include the autosomes, not the sex chromosomes).

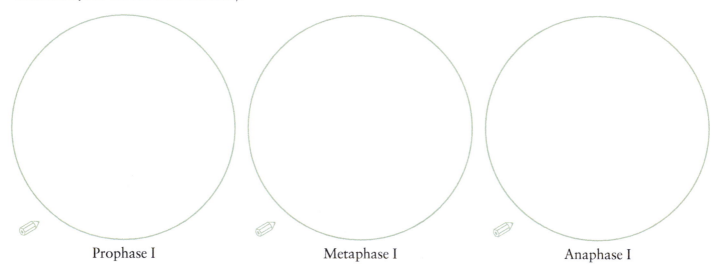

Prophase I   Metaphase I   Anaphase I

3 Illustrate the four products of your meiotic event in the gametes provided.

Materials Required

☐ Coin
☐ Genome template and traits template (included)
☐ Scissors

1 Select two of your gametes from step 3 in Activity 4.

2 Flip a coin to determine which gametes will be those of your mate (or a neighbor, if you are working on this exercise with another student). For example, for each locus, heads is the dominant allele and tails is the recessive allele.

3 Using the second gamete you chose and the gamete you started with, fill out Table 13.3.

   a. Copy the genotype information for "Your Gamete."

   b. Copy the genotype information for your "Neighbor's Gamete."

4 Complete the table by providing the genotype and phenotype information for the offspring you just built.

5 Build your offspring (F1) using the second model face. Be sure to select at least one X gamete. Briefly explain why some traits are the same as your original parent and why some may be totally different from either parent. Why would you need to select at least one X gamete?

TABLE **13.3** Offspring Genome

Genotype		
Your Gamete		
		Sex:
		Hairline:
		Nose:
		Eyes:
		Eyebrows:
		Mouth:
		Earlobes:
Neighbor's Gamete		
		Sex:
		Hairline:
		Nose:
		Eyes:
		Eyebrows:
		Mouth:
		Earlobes:

Genotype		Phenotype
Offspring (F1)		
		Sex:
		Hairline:
		Nose:
		Eyes:
		Eyebrows:
		Mouth:
		Earlobes:

Activity 6 Synthesis

Materials Required
No materials required for this activity

Now let's tie it all together.

1 Genes on the same chromosome may be inherited together. How do linked genes violate Mendel's law of independent assortment?

_____

_____

_____

_____

_____

2 Some English Springer Spaniels suffer from canine phosphofructokinase (PFK) deficiency. These dogs lack an enzyme that is crucial for extracting energy from glucose molecules. Affected pups have extremely weak muscles and die within weeks. A DNA test is available to identify male and female dogs that are carriers. Why would breeders wish to identify carriers if these dogs are not affected?

_____

_____

3 In the given pedigree, determine if the disorder's mode of inheritance is X-linked recessive, autosomal dominant, or autosomal recessive. Explain your reasoning.

_____

_____

_____

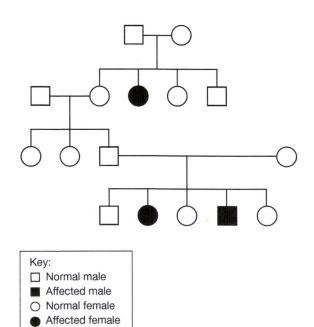

Key:
☐ Normal male
■ Affected male
○ Normal female
● Affected female

Name _____ Date _____ Section _____

Exercise 14
Patterns of Inheritance: Pedigrees and Sex-linked Traits

Objectives

At the completion of this exercise, students will be able to:

1 Define and apply key terms associated with inheritance.

2 Describe how a pedigree is used to analyze the transmission of an inherited trait over the courses of several generations.

3 Infer whether inheritance from a pedigree is autosomal, sex linked, dominant, or recessive.

4 Identify the general probability of inheritance of a specific trait as inferred from a pedigree.

5 Deduce the genotypes and phenotypes of individuals in pedigree charts.

6 Describe the inheritance and expression of human blood groups.

ⓘ Background Information

Pedigree charts are drawn with horizontal lines representing paired couples and vertical lines indicating lines of descendants. Children are listed in birth order from left to right. Females are typically drawn with circles and males with squares. Affected individuals are shaded.

Sex-linked traits are genetically based traits that are carried on sex-specific chromosomes. Usually, they are carried on the X chromosome. This generally results in males being more susceptible to these diseases because they have only one X chromosome. If the allele coding for the trait is present on the male's single X chromosome, the male will show the trait; it does not have another option, as it does not have a homologous allele on another chromosome that could be expressed.

Activity 1 Key Terms

Materials Required
❑ Biology textbook or access to internet resources

Define the following terms:

Pedigree

_____

_____

X chromosome

_____

_____

Y chromosome

_____

_____

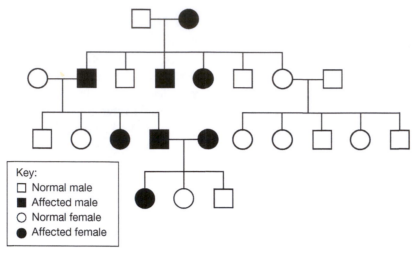

Activity 2 | Inheritance of Sex-Linked Traits

Materials Required

No materials required for this activity

Key:
☐ Normal male
■ Affected male
○ Normal female
● Affected female

FIGURE **14.1** ● Pedigree 1.

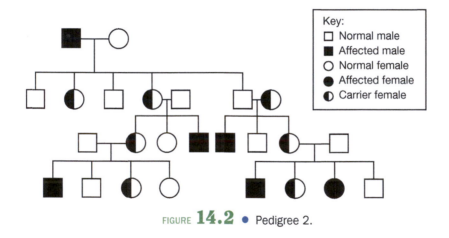

Key:
☐ Normal male
■ Affected male
○ Normal female
● Affected female
◑ Carrier female

FIGURE **14.2** ● Pedigree 2.

1 Which pedigree (Figure 14.1 or Figure 14.2) appears to demonstrate the inheritance of a sex-linked trait, and why?

2 Which pedigree (Figure 14.1 or Figure 14.2) demonstrates the inheritance of an autosomal trait, and why?

3 List some common sex-linked human diseases.

4 List some common autosomal human diseases.

Activity 3 Genetic Pattern of Color Blindness

Materials Required
No materials required for this activity

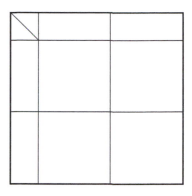

1 Greg is color blind and marries Susan, who is a carrier for color blindness. Color blindness is an X-linked, genetic trait. Draw the Punnett square for their children. Label each individual in the chart with their actual or predicted genotype.

2 Is there any chance they could have a son who is not color blind? If so, what is the probability of Greg and Susan having a son with normal vision?

3 What is the probability of Greg and Susan having a color-blind daughter?

4 Red-green color blindness is inherited as an X-linked recessive trait. How can a man with normal color vision father a daughter who is red-green color blind?

Activity 4 Synthesis

Materials Required
No materials required for this activity

Now let's tie it all together.

1 Draw a 3-generation pedigree chart of your family in the space provided. Use squares for the males in the family and circles for the females. Label each generation with Roman numerals.

2 Look at the traits in Table 14.1 and designate *your* phenotype for each trait (if present) by circling the trait.

3 Choose one of your recessive phenotypes from the chart and track it on your pedigree.

Label each individual in the chart with their expected genotype.

Note: This exercise is a simplification of the actual genetic and environmental mechanisms that eventuate in a person's phenotype. All phenotypes are the result of gene-by-environment interactions. Also note that traits like eye color, hair color, tongue rolling, widow's peak, detached or attached earlobes, hitchhiker's thumb etc., once thought to be single-locus, two-allele traits are more likely due to more than one gene or are strongly influenced by environment. See *Online Mendelian Inheritance in Man* at **http://www.ncbi.nlm.nih.gov/omim** for many more examples.

TABLE **14.1** Dominant and Recessive Alleles of Common Human Traits

Dominant Trait in Humans	Recessive Trait in Humans
A blood type	O blood type
Abundant body hair	Little body hair
Astigmatism	Normal vision
B blood type	O blood type
Baldness (in male)	Not bald
Broad lips	Thin lips
Broad nose	Narrow nose
Dwarfism	Normal growth
Hazel or green eyes	Blue or gray eyes
High blood pressure	Normal blood pressure
Large eyes	Small eyes
Migraine	Normal
Mongolian Fold	No fold in eyes
Nearsightedness	Normal vision
Rh factor (+)	No factor (Rh−)
Second toe longest	First or big toe longest
Short stature	Tall stature
Six fingers	Five fingers normal
Webbed fingers	Normal fingers
Tone deafness	Normal tone hearing
White hair streak	Normal hair coloring

Exercise 15
Genetics of ABO and Rh Blood Groups

Objectives

At the completion of this exercise, students will be able to:

1 Define and apply key terms associated with the genetics of ABO and Rh blood groups.

2 List the phenotype and genotype of possible ABO blood types.

3 Determine compatibility among donors and recipients using the ABO and Rh blood groups.

4 Explain the concepts of universal donor and universal recipient.

5 Discuss the role of co-dominant and recessive alleles in determining blood type.

TABLE **15.1** Rh Inheritance

Rh Factor	Possible Genotypes
Rh+	Rh+/Rh+ **or** Rh+/Rh−
Rh−	Rh−/Rh−

Note: Rh Inheritance is independent of ABO blood type.

ⓘ Background Information

Blood typing is important for many reasons. It is particularly critical for identifying compatible blood transfusion donors and recipients, and in some cases can be used to determine paternity. Some blood types are compatible, meaning that cells of each type of blood can be mixed without serious consequences. Mixing of other types can be lethal. Thus, it is important to understand the distinguishing characteristics of the different blood groups.

There are four major blood types: A, B, AB, and O. Each type is based on the different antigens, designated i, I^A and I^B. I represents the antigen that is attached to blood cells. I^A produces type A blood, I^B produces type B, and i produces type O blood. I^A and I^B are dominant over i, and only ii people have type O blood. I^AI^A or I^Ai have type A blood, while those with I^BI^B or I^Bi have type B. I^AI^B people have both phenotypes, because they express both of their alleles. Therefore, the alleles I^A and I^B are said to be codominant (Fig. 15.1).

Rh factor further discriminates blood types. It is an antigen that is independent of the antigens responsible for blood types. Rh is either present or not present on the surface of blood cells. Thus, blood types can be either Rh positive or Rh negative.

Blood that contains Rh factor (positive) can receive blood that is positive or negative, but those who do not have Rh factor (negative) can only receive blood that is also negative for Rh factor. If a fetus is Rh+ and the mother is Rh−, and a small amount of the baby's blood comes into contact with the mother's blood, then the mother may make antibodies to the Rh antigens in the baby's blood, and would react to the baby as if it were an allergen. Thus, this usually only occurs in subsequent pregnancies of Rh− women where the father is Rh+, leading to a Rh+ pregnancy. In this case, the mother's blood, having made antibodies to Rh+ blood, may cross the placenta and attack the baby's blood like it would any other antigen, breaking down the fetus' blood cells (erythroblastosis foetalis). Thus, during prenatal care, all pregnant Rh− women are provided injections of IgG anti-D antibodies (anti-Rh), which prevents them from developing antibodies against the fetus.

Activity 1 Key Terms

Materials Required
☐ Biology textbook or access to internet resources

Define the following terms:

Antigen

_____

_____

Antibody

_____

_____

Erythroblastosis foetalis

_____

_____

Dominance (draw a picture to illustrate an example)

_____

_____

_____

Incomplete dominance (draw a picture to illustrate an example)

_____

_____

_____

Codominance (draw a picture to illustrate an example)

_____

_____

_____

_____

Rh factor (draw a picture to illustrate an example)

_____

_____

_____

_____

Activity 2 Blood Types

Materials Required
❏ Biology textbook or access to internet resources

1 Use your textbook or internet resources to complete Table 15.2 with diagrams that illustrate the appropriate components.

TABLE **15.2** Components of Blood

	Group A	Group B	Group AB	Group O
Red blood cell type	(diagram: red blood cell labeled A with antigens)			
Antibodies in plasma	(diagram: Anti-B)			
Antigens in red blood cell	(diagram: A antigen)			
Can donate blood to	A and AB			

2 Which blood cell alleles are co-dominant? Recessive?

_____

_____

3 Complete Punnett square A based upon a cross between individuals with the following genotypes: AO & BO.

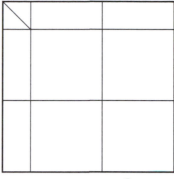

Punnett square **A**

4 What are the genotypes of their offspring? What are the phenotypes of their offspring?

_____

_____

5 Complete Punnett square B based upon a cross between AO and OO blood types (**Hint:** O is recessive).

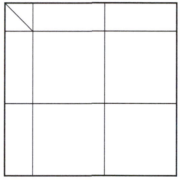

Punnett square **B**

6 What is the probability of one of the offspring having an O phenotype?

7 Complete Punnett square C for an individual with an AO genotype that mates with another AO genotype individual.

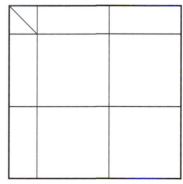

Punnett square **C**

8 What is the probability of an A type phenotype?

9 What is the probability of an OO genotype?

Activity 3 Rh Factor

Materials Required
No materials required for this activity

1 Complete Table 15.3.

TABLE **15.3** Blood Types and Donation

Blood Type	Rh Factor	Can Receive:	Can Donate To:
A−			
A+			
B−			
B+			
AB−			
AB+			
O−			
O+			

2 Genie and Dan have just had a baby girl. Genie's blood type is A− and Dan's is A+.

Complete Punnett square D to reflect the baby girl having an O phenotype.

3 Maria (AA−) and Jason (AB−) Phillips had twins, a boy and a girl. Lily (AO−) and Charles (AO−) Smith had a boy. When Maria received her son's birth certificate, his blood type was labeled O−. Based on the information provided, what is the best possible explanation for the O− blood type on the birth certificate Maria received? What is the best explanation for the Smiths' son whose blood was labeled AB−?

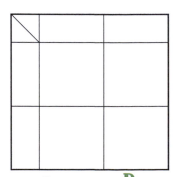

Punnett square **D**

Activity 4 Synthesis

Materials Required
No materials required for this activity

Now let's tie it all together.

1 A type AB father and a type A (AA or AO) mother are having a baby. Use the space provided to generate Punnett squares to assist you in answering the following questions:

   a. If the mother is AA, what is the probability that the baby has type A blood? Type B? Type AB? Type O?

_____

_____

_____

_____

_____

_____

generate Punnett square here

   b. If the mother is AO, what is the probability that the baby has type A blood? Type B? Type AB? Type O?

_____

_____

_____

_____

_____

generate Punnett square here

2 The mother of a child is blood type O + and the child is A −.

   a. What are the possible blood types of the father? Complete Punnett square E to determine your answer.

_____

   b. If the Rh factor of a child is negative, does that mean that one of the parents has to be negative?

_____

   c. Annotate the Punnett square you just made to show how you determined your answer.

Punnett square **E**

3 Assume that both parents of a child are Rh−.

   a. Could two Rh− parents give birth to a Rh+ child?

_____

   b. Could two Rh+ parents give birth to a Rh− child?

_____

4 Complete Table 15.4, which summarizes the heritability of various blood groups.

DAD

TABLE **15.4** Blood Group Inheritance

Blood type	Genotype	O — ii (OO)		A — Iᴬi (AO)		A — Iᴬ Iᴬ (AA)		B — Iᴮi (BO)		B — Iᴮ Iᴮ		AB — Iᴬ Iᴮ		Phenotype(s)/ Genotype
O	ii (OO)											A or B		Phenotype(s)
												AO	BO	Genotype
												AO	BO	
A	Iᴬi (AO)					A								Phenotype(s)
						AA	AA							Genotype
						AO	AO							
A	Iᴬ Iᴬ (AA)	O or A												Phenotype(s)
		AO	AO											Genotype
		OO	OO											
B	Iᴮi (BO)													Phenotype(s)
														Genotype
B	Iᴮ Iᴮ (BB)													Phenotype(s)
														Genotype
AB	Iᴬ Iᴮ (AB)							A or B						Phenotype(s)
							AB	AO						Genotype
							BB	BO						

(MOM — blood types listed down the left side)

Genetics: Molecular

· · · · · · · · ·

In This Section

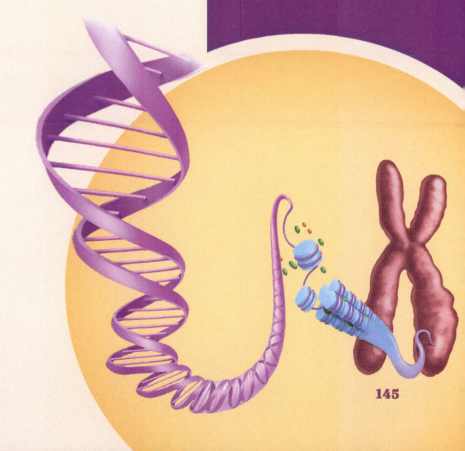

Exercise 16
Structure and Replication of DNA

Objectives

At the completion of this exercise, students will be able to:

1 Define and apply key terms associated with the structure and replication of DNA.

2 Describe Watson-Crick base pairing and its importance for the structure of DNA.

3 Illustrate components of DNA structure, including the number of strands, polarity (5' to 3'), complementary strands, anti-parallel strands, and primary and secondary structure.

4 Explain the importance of the strong, covalent bonds of the phosphodiester (sugar-phosphate) backbone and the weaker hydrogen bonds that form between the nitrogenous bases.

5 Explain the fundamental steps involved in DNA replication.

6 Distinguish between the DNA template strand and the newly synthesized strand.

7 Discuss the necessity of directionality in DNA synthesis (5' to 3').

8 Compare and contrast the leading and lagging strands of new synthesized DNA.

9 Illustrate the role of each enzyme or protein required in the process of DNA synthesis (at the replication fork).

ⓘ Background Information

DNA was not always considered the best candidate for storing genetic data. Many researchers felt that genetic information would more likely be stored and transmitted as proteins, because proteins are composed of 20 amino acids, which could possibly code for many more unique traits than only the four nucleotides that make up DNA (adenine, guanine, thymine and cytosine). But clever experiments by Avery, MacLeod, and McCarty (1944); Chargaff (1949); and Hershey and Chase (1952) provided convincing evidence that the blueprint for life was coded for in DNA. In 1953, James Watson and Francis Crick used Rosalind Franklin's experimental data (allegedly without her permission), along with some of Maurice Wilkins' data, to propose a double helical model for DNA. Their discovery was interesting, not so much because they cracked the structure of DNA, but because, as they point out at the very end of their paper, "It has not escaped our notice that the specific pairing we have postulated immediately suggests a possible copying mechanism for the genetic material." Thus, understanding the structure of DNA provided insight into mechanisms of replication and the propagation of life.

Activity 1 Key Terms

Materials Required
❑ Biology textbook or access to internet resources

Define the following terms:

DNA

_____

_____

Helix

_____

_____

Purine

Pyrimidines

Base pair

dNTP

Nucleotide

Polynucleotide

Nucleoside

Deoxyribose

Phosphodiester bond

Strand directionality

Antiparallel

Primary structure of DNA

_____

_____

Secondary structure of DNA

_____

_____

DNA polymerase

_____

_____

Major groove

_____

_____

Minor groove

_____

_____

Dehydration synthesis reaction

_____

_____

Activity 2 Structure of DNA

Materials Required
☐ Colored pencils or marking pens

1 Color and label the following structure of a deoxyribonucleotide (Fig. 16.1). Note that the nitrogen-containing bases could be adenine (A), thymine (T), guanine (G), or cytosine (C).

2 Color the phosphate group, the sugar (deoxyribose), and the nitrogenous base. Identify the 5' carbon and the 3' carbon.

3 Highlight the carbon atom that is attached to the phosphate group. This is the 5' carbon.

4 Highlight the carbon atom that is attached to the hydroxyl (–OH) group. This is the 3' carbon.

CH₂

O

OH

FIGURE **16.1** ● A deoxyribonucleotide.

5 Now look at the primary structure of DNA (Fig. 16.2). Color and label the diagram of the primary (1°) structure of DNA. Make special note of the pyramidines and purines, and how the sugar phosphate backbone of the DNA strand is linked together.

6 Highlight the phosphodiester bonds that link the deoxyribonucleotides.

7 Designate the directionality of each strand (5' to 3') in Figure 16.2 with an arrow.

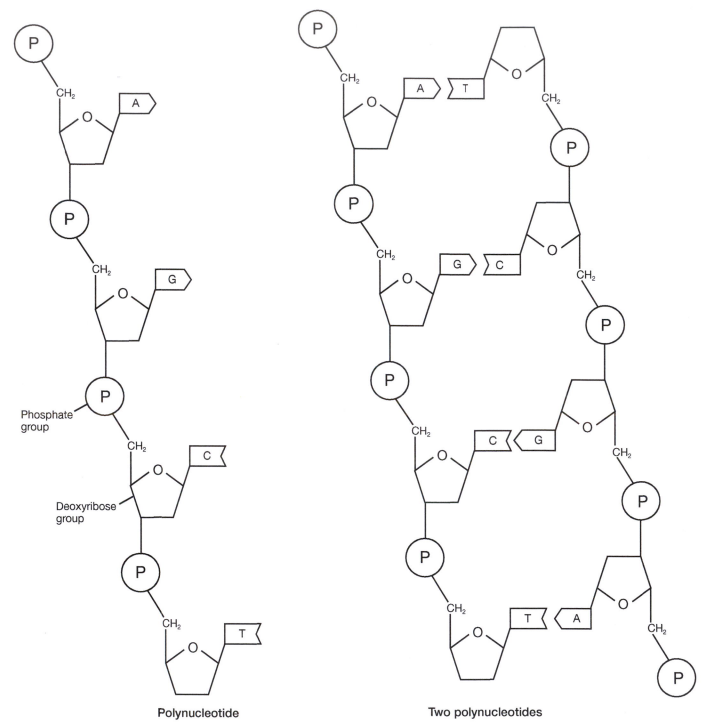

Polynucleotide   Two polynucleotides

FIGURE **16.2** ● Primary structure of single-stranded and double-stranded DNA.

8 Now look at the secondary structure of DNA (Fig. 16.3). Color and label the diagram of the secondary structure of DNA. Be sure to designate the sugar phosphate "backbone," complementary base pairs, and the number of hydrogen bonds that hold them together.

9 Label the antiparallel strands and their directionality (polarity; their 5' to 3' ends run in opposite directions).

10 Label the major and minor groove.

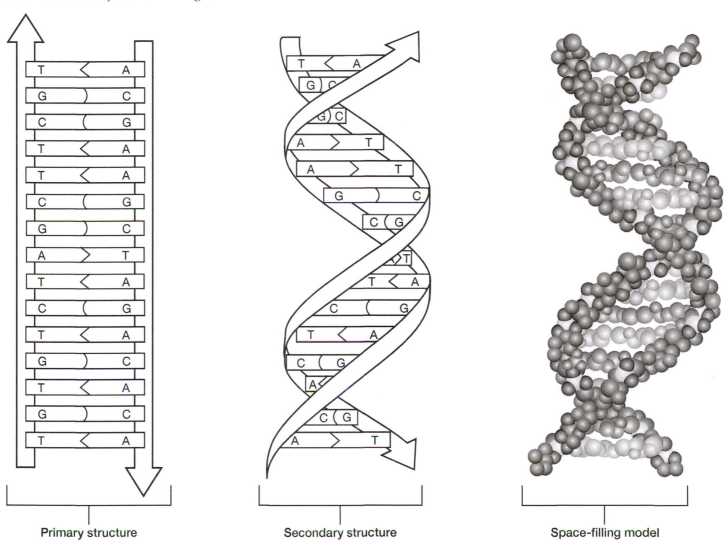

Primary structure   Secondary structure   Space-filling model

FIGURE **16.3** ● Primary, secondary, and a space-filling model of double-stranded DNA.

Activity 3 DNA Synthesis

Materials Required
☐ Colored pencils or marking pens

1 Annotate Figure 16.4.

    a. Highlight the directionality (polarity) of each strand.

    b. Distinguish the template strand from the growing strand.

    c. Identify the activated nucleotide (dNTP) being added to the growing strand.

    d. Highlight the formation of the phosphodiester bond that forms from the dehydration synthesis reaction.

    e. Note that the enzyme DNA polymerase catalyzes this reaction. The activated dNTP has three phosphorus groups at the 5' carbon that will bond to the 3' carbon on the growing strand.

    f. Note the products of the reaction: elongated DNA, two phosphates, and water.

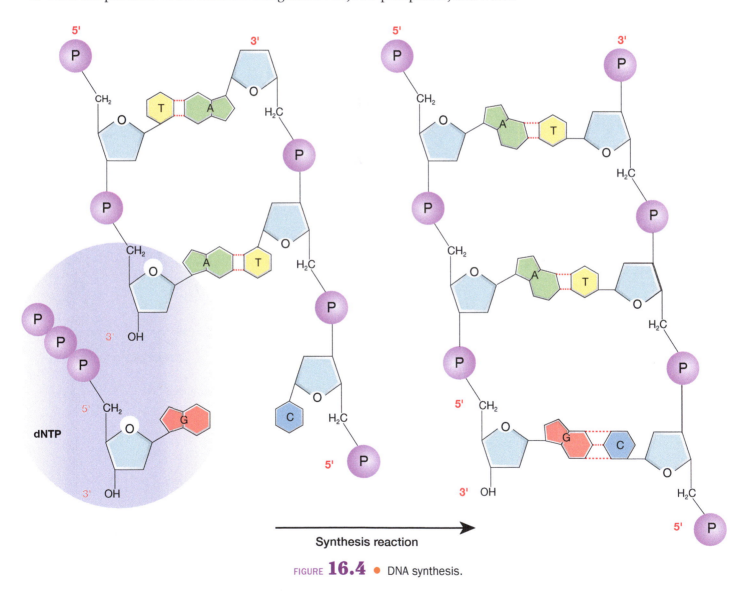

FIGURE **16.4** ● DNA synthesis.

Activity 4 The Replisome

Materials Required
☐ Colored pencils or marking pens

1 Annotate Figure 16.5, which is a schematic of the replication fork.

    a. In each box, write the name of the player and its function. Include: DNA polymerase III, sliding clamp, helicase, single-stranded binding protein (SSBPs), topoisomerase, RNA primer, newly synthesized DNA, primase, and DNA ligase.

    b. Identify the leading strand and the lagging strand of newly synthesized DNA and their directionality (5' to 3') of synthesis.

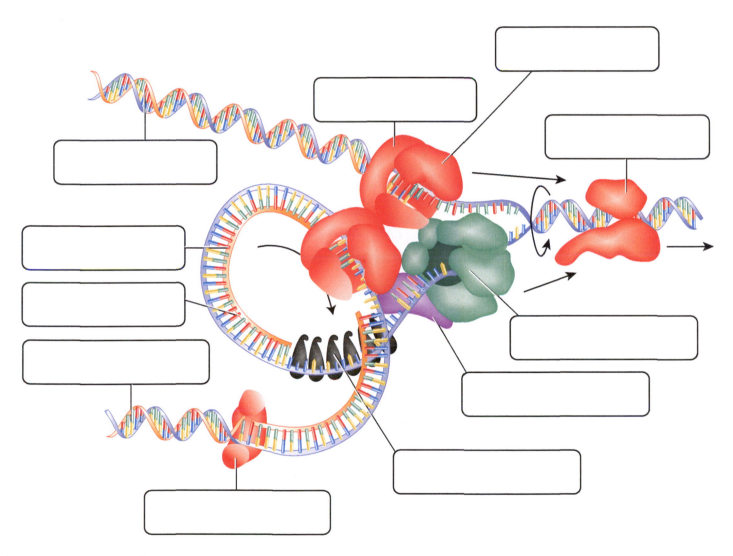

FIGURE **16.5** ● The replisome, which consists of the replication fork and all the major players involved in DNA synthesis.

Activity 5 Synthesis

Materials Required
No materials required for this activity

Now let's tie it all together.

1 What is the function of DNA?

_____

_____

2 What is a monomer unit of DNA?

_____

_____

3 Name the bases that are purines and those that are pyramidines.

_____

4 Which complementary base pairs are bonded together the strongest (share the most hydrogen bonds)?

_____

5 Which complementary base pairs are most easily denatured (broken, resulting in separation of the base pairs)?

_____

6 Compare and contrast a nucleoside and a nucleotide.

_____

_____

7 What are the products of a dehydration synthesis reaction?

_____

8 Why can't both strands be synthesized continuously? Or, in other words, why must there be a leading and a lagging strand?

_____

_____

9 Which are the three components of the replication fork that are only involved in lagging strand synthesis?

_____

_____

10 Why can DNA never be synthesized in the 3' to 5' direction?

_____

_____

_____

Exercise 17
Structure and Function of RNA

Objectives:

Students will be able to:

1 Define and apply key terms associated with the structure and function of RNA.

2 Illustrate the structure of RNA.

3 Differentiate among three major types of RNA and their roles in transcription.

4 Describe the role of each type of RNA in protein synthesis.

5 Explain the genetic code and its relationship between DNA, RNA, and protein.

ⓘ Background Information

As a single-stranded polynucleotide, RNA has numerous levels of structure and functioning. For example, double-stranded DNA can have secondary structure (an alpha helix), but because RNA is single-stranded it can form hairpins and loops, as well as distinctive three-dimensional shapes or tertiary structures. It can even form associations with other RNA molecules, or quaternary structures. Similarly, there are different types of RNA that serve different functions in the cell. Among the most important are mRNA, rRNA, and tRNA.

Activity 1 Key Terms

Materials Required
❏ Biology textbook or access to internet resources

Define the following terms:

RNA

_____

_____

tRNA

_____

_____

Aminoacyl tRNA

_____

_____

mRNA

_____

_____

rRNA

Ribosome

Large ribosomal subunit

Small ribosomal subunit

Codon

Anticodon

Anticodon loop

Amino acid

Genetic code

Activity 2

Structural Differences between DNA and RNA

Materials Required

No materials required for this activity

1 Color and label the DNA and RNA strands in Figure 17.1. Include labels for nucleotide bases, base pairs, phosphodiester backbone, adenine, cytosine, guanine, thymine, and uracil.

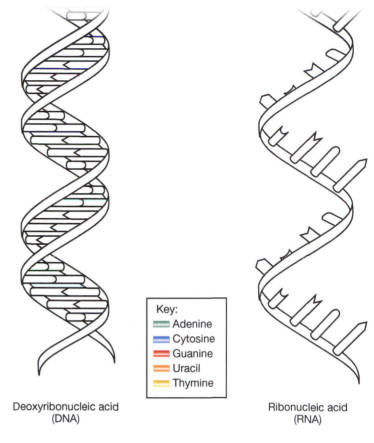

Key:
▬ Adenine
▬ Cytosine
▬ Guanine
▬ Uracil
▬ Thymine

Deoxyribonucleic acid
(DNA)

Ribonucleic acid
(RNA)

FIGURE **17.1** ● Comparison of RNA and DNA.

Activity 3 Structure of tRNA

Materials Required

No materials required for this activity

1 Label the tRNA molecule in Figure 17.2. Be sure to identify the aminoacyl attachment site (the site where the aminoacyl synthetase connects to it and links it to its correct amino acid), the acceptor arm, the directionality of the anticodon loop (5' to 3'), and the complementary mRNA sequence recognized by the anticodon loop (and its directionality).

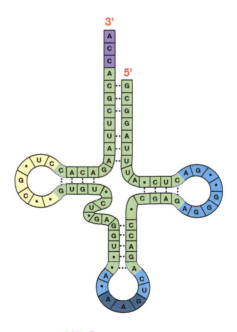

FIGURE **17.2** ● Structure of tRNA.

Activity 4 Structure of rRNA

Materials Required
No materials required for this activity

rRNA is the predominant and most important functional component of the ribosome, catalyzing the peptide bond that joins the two adjacent amino acids that are held by the ribosome.

1 Label Figure 17.3. Be sure to include mRNA, rRNA, small ribosomal subunit, large ribosomal subunit, A site (binding site for the aminoacylated tRNA), P site (binding site for the peptidyl tRNA), and E site (binding site for the tRNA that is to be released).

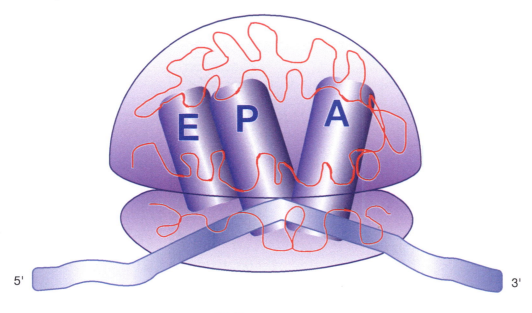

5'   3'

FIGURE **17.3** ● rRNA and the ribosome.

 **Activity 5** Synthesis

Materials Required
No materials required for this activity

Now let's tie it all together.

1 What are the major differences between RNA and DNA?

_____

_____

2 Transcribe the following DNA sequence into its mRNA:

5- GAG CTA GTG ATA AGC CTC ATC GTG GAG TCA -3

_____

_____

3 Using the genetic code (Figure 17.4), translate the following mRNA sequence into the amino acid sequence that it codes for:

5- GAA CUA GUG AUC UCG CUA AUU GUA GAG UCC -3

First Base	Second Base				Third Base
	U	C	A	G	
U	UUU phenylalanine	UCU serine	UAU tyrosine	UGU cysteine	U
	UUC phenylalanine	UCC serine	UAC tyrosine	UGC cysteine	C
	UUA leucine	UCA serine	UAA stop	UGA stop	A
	UUG leucine	UCG serine	UAG stop	UGG tryptophan	G
C	CUU leucine	CCU proline	CAU histidine	CGU arginine	U
	CUC leucine	CCC proline	CAC histidine	CGC arginine	C
	CUA leucine	CCA proline	CAA glutamine	CGA arginine	A
	CUG leucine	CCG proline	CAG glutamine	CGG arginine	G
A	AUU isoleucine	ACU threonine	AAU asparagine	AGU serine	U
	AUC isoleucine	ACC threonine	AAC asparagine	AGC serine	C
	AUA isoleucine	ACA threonine	AAA lysine	AGA arginine	A
	AUG (start) methionine	ACG threonine	AAG lysine	AGG arginine	G
G	GUU valine	GCU alanine	GAU aspartic acid	GGU glycine	U
	GUC valine	GCC alanine	GAC aspartic acid	GGC glycine	C
	GUA valine	GCA alanine	GAA glutamic acid	GGA glycine	A
	GUG valine	GCG alanine	GAG glutamic acid	GGG glycine	G

FIGURE **17.4** ● The universal genetic code (RNA).

4 How would the effect of a mutation in a copy of a gene's DNA on the phenotype of an organism be different than a mutation in a copy of a gene's RNA?

5 What amino acid will the tRNA in Figure 17.2 carry to the ribosome?

_____

_____

6 In what ways does RNA have a more diverse function than DNA?

_____

_____

_____

7 Why is DNA better at storing information than RNA?

_____

_____

Exercise 18
Transcription and Translation

Objectives

At the completion of this exercise, students will be able to:

1 Compare and contrast the components, location, and function of DNA, mRNA, rRNA, and tRNA.

2 Derive mRNA sequences from DNA sequences (transcription).

3 Derive amino acid sequences from mRNA (translation).

4 Outline the processes of transcription and translation and where these processes take place in the cell.

5 Describe the functions of the basic components needed to successfully carry out transcription and translation.

6 Learn the processes of transcription and translation by teaching it to others.

DNA can also transcribe RNA that functions as ribosomal RNA (rRNA), transfer RNA (tRNA), or other non-coding or enzymatic RNAs. If the gene that was transcribed codes for a protein, then the mRNA will serve as the template RNA strand for synthesizing the protein by way of a process called *translation*. Translation is the principal step in protein synthesis. The process of translation uses cellular ribosomes to read the mRNA, assemble the amino acids coded for in the mRNA, and join the amino acids into a protein, a polypeptide, by catalyzing peptide bond formation between the amino acids. Together, transcription and translation form the basis of gene expression, which is the process by which genetic information is used to generate a functional gene product.

(i) Background Information

Transcription is the principal step in gene expression and occurs when RNA polymerase copies the information on a segment of DNA (usually a gene) into a strand of message RNA (mRNA), which typically codes for a protein. But

Activity 1 Key Terms

Materials Required
❏ Biology textbook or access to internet resources

Define the following terms:

Transcription bubble

_____

_____

RNA polymerase

_____

_____

Sense strand (coding strand; nontranscribed strand)

Antisense strand (template strand; transcribed strand)

Nuclear pore

Ribosome

Initiation

Elongation

Termination

Cytoplasm

Activity 2 Transcription

Materials Required
❑ Colored pencils, pens, or markers

1 In Figure 18.1, color and label the key players and processes involved in transcription. Be sure to label RNA polymerase, template strand of DNA, transcription bubble, complementary non-template strand, ribonucleotides, and RNA (pre-mRNA).

2 Label the 5' and 3' ends of each polynucleotide (nucleic acid molecule) and the direction of RNA synthesis (transcription).

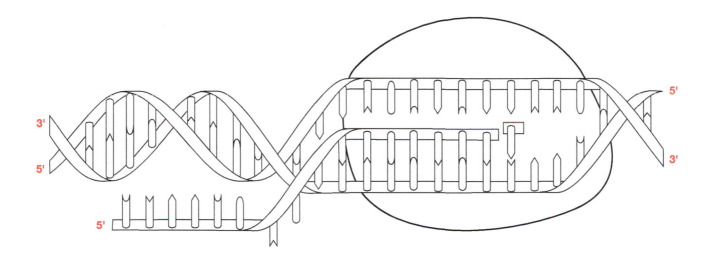

FIGURE **18.1** ● Transcription.

Activity 3 Translation

1 In Figure 18.2, color and label the key players and processes involved in *translation*. Be sure to label ribosome, incoming tRNA charged with its amino acid aminoacyl tRNA, mRNA, large ribosomal subunit, small ribosomal subunit, aminoacyl tRNA synthetase, and growing peptide chain (polypeptide).

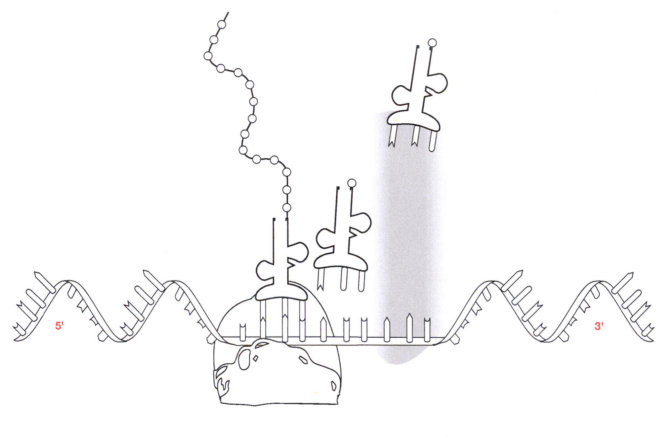

FIGURE **18.2** ● Translation.

Activity 4 Transcription and Translation Chart

1 Complete the missing information in Table 18.1. Column 1 contains the bases that complement the coding strand. A common convention is to read them from left to right, but in this exercise we use 5' to 3' notation and read from top (5') to bottom (3'). Column 2 contains the template, or coding strand. Column 3 contains the mRNA. Column 4 contains the sequence of the tRNA anticodon loop. Column 5 contains the amino acid coded for by the mRNA (you will need a table of the genetic code to determine which amino acid goes here). Note that just like a ribosome you will need to "scan" the mRNA sequence for the start site. The answers in the first row are provided.

TABLE **18.1** Coding for Transcription and Translation

DNA		mRNA	tRNA anticodon loop	Amino acid
Complementary Strand	**Template Strand**			
G⁵'	C³'	G⁵'	C	
G				
T				
C				
T				
A				
T				
G				
T				
C				
A				
G				
G				
C				
C				
A				
C				
C				
T				
G				
C				
C				
G				
C				
G				
G				
A				
C				
T³'				
Complement	Gene/template			

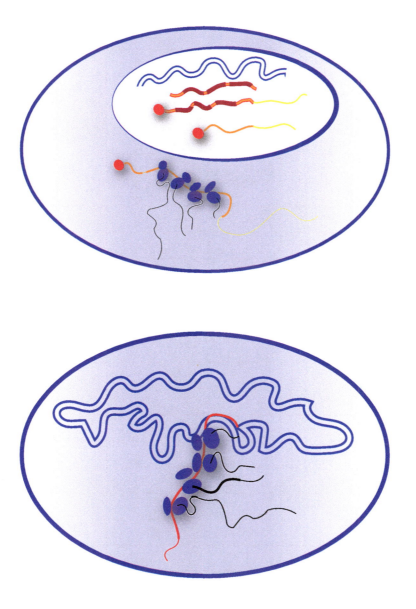

Activity 5 Comparison of Prokaryotic and Eukaryotic Transcription and Translation

Materials Required
❑ Colored pencils, pens, or markers

Because prokaryotes lack a nuclear membrane, genes can be transcribed and translated simultaneously. In eukaryotes, RNA is modified in the nucleus before it is shuttled into the cytoplasm to be translated. Transcribed eukaryotic pre-mRNA is first capped on the 5' end and a poly-A tail is added to the 3' end. Next, a process called splicing removes the introns and joins the exons together. The final product, a mature mRNA, is then shuttled out of the nucleus to be translated.

1 Label Figure 18.3, which illustrates the primary differences between prokaryotic and eukaryotic transcription. Labels should include plasma membrane, nuclear membrane, ribosomes, DNA, pre-mRNA, mature mRNA, mRNA, intron, exon, 5' cap, 3' poly-A tail, and protein.

FIGURE **18.3** • Differences between prokaryotic and eukaryotic transcription.

Activity 6 Teaching Transcription

Materials Required
❑ Poster board
❑ Colored pencils, pens, or markers
❑ Biology textbook or access to internet resources

1 On one side of your poster board, draw a schematic diagram illustrating the process of transcription. In creating your teaching aids you may use online or textbook resources as a guide.

2 Teach the processes to five different people and have them initial your poster. Be sure to include in your poster and discussions the following topics.

  a. Where does transcription occur in prokaryotes? Where does it occur in eukaryotes?

  b. Describe the three phases of transcription and the players (proteins and DNA features) that are involved in each phase.

  c. Discuss the directional nature of the process.

  d. What is the product of transcription? What was used to create that product?

  e. Describe RNA processing and the difference between pre-mRNA and mRNA.

Activity 7 Teaching Translation

Materials Required
❑ Poster board
❑ Colored pencils, pens, or markers
❑ Biology textbook or access to internet resources

1 On the other side of your poster board, draw a schematic diagram illustrating the process of translation. In creating your teaching aids you may use online or textbook resources as a guide.

2 Teach the processes to five different people and have them initial your poster. Be sure to include in your poster and discussions the following topics.

  a. Where does translation occur?

  b. What is the relationship between the codon in DNA, the codon in mRNA, and the tRNA anticodon?

  c. Describe the ribosomal structure and the purpose of the "E," "P," and "A" sites.

  d. What is the product of translation? What reagents are used to create this product?

Activity 8 Synthesis

Materials Required

No materials required for this activity

Now let's tie it all together.

1 Compare and contrast the structure, location, and function of four key nucleic acids in Table 18.2.

TABLE **18.2** Structures, Locations, and Function of the Nucleic Acids

Nucleic Acid	DNA	mRNA	tRNA	rRNA
Name of sugar in the nucleotide				
Name of the nucleotide bases				
Function of the nucleic acid				
Strandedness (how many strands?)				
Typical length of each molecule				
Location in the cell				

2 A common misconception that students have about translation is that amino acids are produced by translation. What is the correct relationship between amino acids and protein synthesis?

_____

_____

_____

_____

3 One of the most confusing aspects of determining the protein sequence that will be derived from a gene is keeping track of the polarity or directionality of each nucleotide sequence strand. Which DNA strand is identical to the mRNA transcript? Which DNA strand codes the RNA transcript?

_____

_____

4 The words "transcribe" and "translate" are more commonly associated with language and dialogue. For example, a court reporter may transcribe what was said during a trial, or a popular book may be translated from Spanish into English. Why are "transcription" and "translation" such good descriptors for these processes in molecular biology?

_____

_____

_____

Exercise 19
Control of Gene Expression

Objectives

At the completion of this exercise, students will be able to:

1 Define and apply key terms associated with gene expression.

2 Describe how and why genes are regulated.

3 Distinguish between positive and negative mechanisms of gene regulation.

4 Describe the components of the *lac* operon and their role in gene expression.

5 Discuss how mutations to binding sites, repressors, or activators can lead to changes in gene expression.

ⓘ Background Information

Your brain cells contain a copy of every gene your body needs to function properly. But do you really want your brain cells to make insulin? In order for each cell in your body to make the correct gene product, at the right time, and in the right place, cells must carefully regulate the expression of their genes. In bacteria and some eukaryotes, an entire cluster of genes can be under the control of a single promoter. In *E. coli*, the set of genes that regulate how lactose is taken up and used by the cell is called the *lac* operon. The *lac* operon consists of a promoter, an operator, three genes (*lacZ*, *lacY*, and *lacA*), and a terminator.

When lactose is abundant in its environment, and levels of glucose inside the cell are low, it makes sense for an *E. coli* cell to try to use lactose as an energy source. In order to do this, the cell must produce enzymes that can allow lactose into the cell, break lactose down into glucose and galactose, and detoxify the byproducts of its metabolism (the functions of *lacY*, *lacZ*, and *lacA*, genes respectively). However, when glucose is already abundant, or lactose is not present, it would be wasteful (energetically costly) for the cell to make a bunch of enzymes it cannot use. Thus, there should be strong selection against cells that do not control the expression of these genes, while cells that regulate them efficiently should be favored.

Activity 1 Key Terms

Materials Required
❏ Biology textbook or access to internet resources

Define the following terms:

Transcriptional initiation

_____

_____

RNA processing/Post-transcriptional modification

_____

_____

mRNA degradation

_____

_____

Inducers

_____

_____

Repressors

_____

_____

Operators

_____

_____

Activators

_____

_____

Enhancers

_____

_____

Silencers

_____

_____

Enzymes

_____

_____

Promotors

_____

_____

Upregulation

_____

_____

Downregulation

_____

_____

Introns

_____

_____

Exons

_____

_____

Histones

_____

_____

CAP (cAMP receptor protein)

_____

_____

Negative control

_____

_____

Positive control

_____

_____

Constitutive gene

_____

_____

Facultative gene

_____

_____

Inducible gene

_____

_____

Materials Required

❑ Biology textbook or access to internet resources
❑ Colored pencils

1 Using your textbook or internet resources, label the control elements and genes of the *lac* operon on the template strand of DNA in Figure 19.1. Fill in the appropriate terms in each box: CAP binding site, promotor, operator, *lacZ* gene, *lacY* gene, and *lacA* gene.

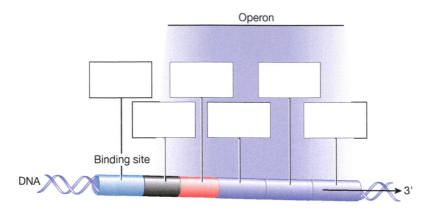

FIGURE **19.1** ● *Lac* operon on the template strand of DNA.

2 What binds to the CAP binding site on the template strand of DNA?

_____

_____

3 What binds to the promotor on the template strand of DNA?

_____

_____

4 What binds to the operator on the template strand of DNA?

_____

_____

_____

Materials Required

No materials required for this activity

1 In Figure 19.2, identify which sugar must be present based on the binding site occupancy in each scenario:

Name ____________________ Date ____________ Section ____________

Glucose	Lactose
________	________
________	________
________	________
________	________

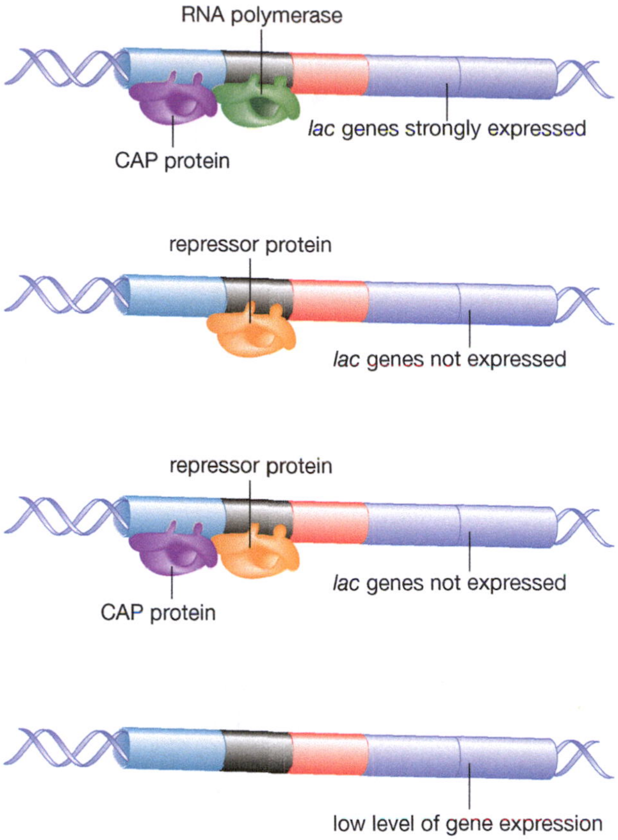

RNA polymerase
CAP protein
lac genes strongly expressed
repressor protein
lac genes not expressed
repressor protein
CAP protein
lac genes not expressed
low level of gene expression

FIGURE **19.2** • Gene regulation.

Activity 4 Regulation of Gene Expression

Materials Required

No materials required for this activity

Whether a gene is under positive or negative control depends on how the gene is expressed when *no* regulatory protein is present. Genes under positive control are actively expressed when an activator protein is present (Fig. 19.3). Genes under negative control are actively expressed *unless* a repressor protein is present (Fig. 19.4).

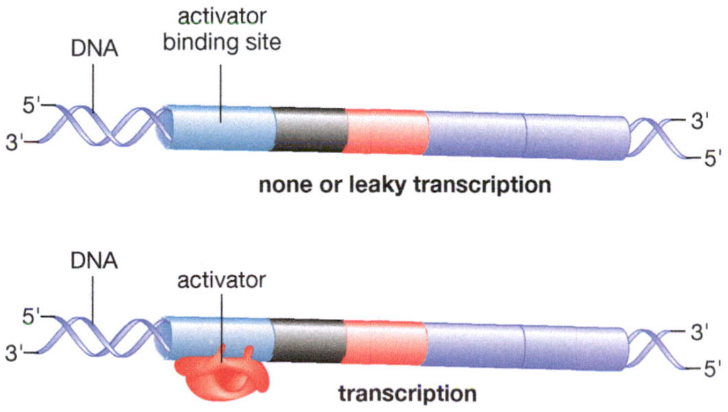

DNA
activator binding site
5'
3'
3'
5'
none or leaky transcription
DNA
activator
5'
3'
3'
5'
transcription

FIGURE **19.3** • Genes under positive control.

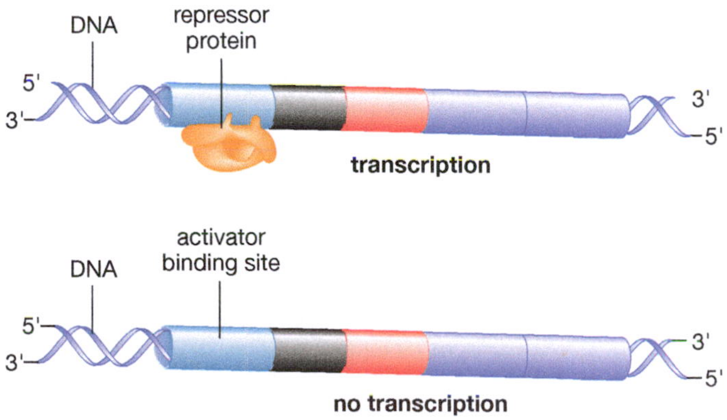

DNA
repressor protein
5'
3'
3'
5'
transcription
DNA
activator binding site
5'
3'
3'
5'
no transcription

FIGURE **19.4** • Genes under negative control.

1 Complete the missing information in columns 2, 3, and 4 of Table 19.1.

TABLE **19.1** Differences between Positive and Negative Control

	Type of Regulatory Protein Involved	Gene Expression when Regulatory Protein Present (on or off?)	What happens if you mutate the regulatory gene (silencer or enhancer?) (expression on or off?)
Positive Control			
Negative Control			

Activity 5 Positive and Negative Regulation of Gene Expression in the *lac* Operon

Materials Required
❏ Colored pencils

The *lac* operon is under negative regulation—the genes in the *lac* operon are always repressed unless a signal molecule comes along and removes the repressor protein. In this case, if lactose is not available to the cell, the repressor binds tightly, and the RNA polymerase cannot bind to the promoter. But when lactose is available, allolactose (a lactose metabolite) acts as a ligand and binds to the repressor protein, causing it to lose its ability to bind to the promoter. The presence and action of ligands on repressor and activator proteins is responsible for switching gene expression on and off.

1 Using the blank space provided after Figure 19.5, illustrate a reconstruction of this figure.

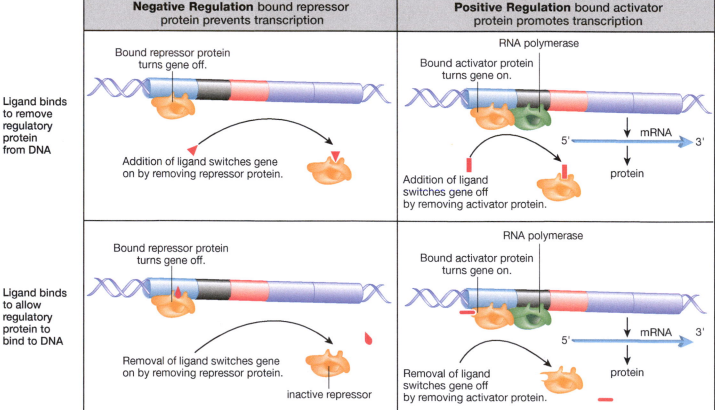

FIGURE **19.5** ● Negative and positive regulation.

Activity 6 Synthesis

Materials Required
No materials required for this activity

Now let's tie it all together.

1 Describe how the activators and repressors interact with the DNA and with each other.

_____

_____

_____

_____

2 Predict the phenotype of the following loss of function mutations in the *lac* operon:

a. Promoter

_____

_____

b. Operator

_____

c. *lacZ*

_____

d. *lacY*

_____

e. *lacA*

_____

f. Terminator

_____

_____

3 Draw an imaginary gene that is regulated by both activators and repressors.

4 Consider the process of gene expression, from transcription through translation:

If translation (protein synthesis) were carried out experimentally under *in vitro* conditions using mRNA from penguin muscle cells, ribosomes from bacterial cells, tRNAs from duck eggs, and amino acids from nematode worms, what products would result?

_____

_____

_____

_____

Exercise 20
Techniques of Molecular Genetics: DNA Extraction

Objectives

At the completion of this exercise, students will be able to:

1 Describe nucleic acid chemistry and extraction of DNA from tissue.

2 Explain where DNA is contained and how it can be extracted.

3 Observe naked DNA.

4 Discuss the implications of recombinant DNA and genetically modified organisms.

extraction process. The contents of each cell are surrounded by a cell membrane that is composed of two layers of phospholipids (phospholipid bilayer). Each phospholipid has a hydrophilic head and two hydrophobic tails. The lipid bilayer is the barrier that keeps proteins, nucleic acids, and other molecules inside the cell and protects it from its surroundings. To isolate DNA from the other components in the cell, the cell membrane must be disrupted, and the DNA must be separated from the many proteins, fats, and sugars that make up the cell.

(i) Background Information

DNA is the storage molecule of genetic information that determines an organism's genotype. It resides inside the cells of all organisms and can be extracted using a simple

Activity 1 Key Term

Materials Required
❑ Biology textbook or access to internet resources

Define the following term:

Recombinant DNA

_____

_____

Activity 2 DNA Extraction

Materials Required
- ❑ 1 resealable plastic bag (sandwich size)
- ❑ 2 strawberries (frozen is OK, but they must be thawed) or ½ banana
- ❑ 2 tsp. dish detergent
- ❑ 1 tsp. of salt
- ❑ ½ cup of water
- ❑ 2 plastic cups
- ❑ 1 coffee filter (or cheesecloth)
- ❑ ½ cup ice cold rubbing alcohol
- ❑ 1 wooden coffee stirrer/toothpick/stick

Setup
In this activity you will extract DNA from living material. Before you begin, here are some helpful suggestions.

1 Make the extraction buffer first and keep it in the refrigerator until you are ready to use it. To make the extraction buffer, add 2 teaspoons of detergent, 1 teaspoon of salt, and ½ cup of water to one of the plastic cups. Stir well, and refrigerate or let stand in an ice bucket.

2 Keep the alcohol ice cold until you are ready to use it.

3 Try to move through each step of the activity as efficiently as possible. In addition to DNA, smashed-up fruit also contains enzymes that chop up DNA if it sits at warm temperatures for too long.

Procedure

1 Remove any of the green stems or leaves on the strawberry (if using a banana, remove the peel).

2 Place fruit into the plastic bag. Seal the bag and gently mash up the fruit with your hands for two minutes. What does this mechanical action do to the cells and the DNA?

3 Make your DNA extraction buffer. Add 2 teaspoons of detergent, 1 teaspoon of salt, and ½ cup of water to one of the plastic cups. Stir well.

4 Open the bag of fruit and add 2 teaspoons of your DNA extraction buffer. What are cell membranes made of? Which parts are hydrophobic and which are hydrophilic?

5 What does soap do to cell membranes? Why would this be helpful in extracting DNA?

6 What function does the salt perform in the solution?

7 Reseal the bag. Making as few soap bubbles as possible, gently mash the fruit for another minute.

8 Place the coffee filter inside the second plastic cup. Open the bag of mashed-up fruit and pour it into the coffee filter. Gently squeeze the remaining liquid into the cup.

9 Add rubbing alcohol to the cup of extracted fruit liquid. To do this, carefully pour down the edge of the cup the same amount of ice cold rubbing alcohol as there is liquid. This separates the DNA from the rest of the fruit cellular debris.

10 Observe the cloudy, white substance that forms on top of your fruit extract.

DNA is not soluble in alcohol (especially if it is very cold), so where is the DNA?

11 Tilt the cup 45 degrees. Collect the DNA using the plastic coffee stirrer/wooden stick/spoon. Why can you see big globs of DNA now, but you don't see them when you're eating your food?

12 Why is it important for scientists to be able to isolate DNA?

13 If you eat DNA all the time, why aren't you harmed by it? What implication does this have on the commercial properties of DNA in your hair conditioner, nutritional supplements, skin cream, or genetically modified plants and animals?

14 Strawberries are octoploid. Bananas are triploid. Kiwi fruits are diploid. How does ploidy level influence the amount of DNA in each cell?

Activity 3 Synthesis

Now let's tie it all together.

1. Genetically altered bacteria can produce human protein products, such as hormones and enzymes that are important for improving human health. Provide an example of one of these gene products and its impact on human health.

2. DNA fingerprinting involves analysis of highly polymorphic DNA sequences that are unique to each individual. Provide an example of an application and its impact on individual humans and society.

3. Transgenic organisms have foreign DNA incorporated into their genomes that can be expressed to improve the properties of agricultural plants and animals. Provide an application and its potential for global impact.

4. Transgenic organisms are only possible because widely different organisms share a common mechanism for DNA replication, transcription, and translation. What are the implications of this on their evolutionary origin?

Evolution
and Origin of Species

• • • • • • • • •

In This Section

Exercise 21
Chemosynthetic Origin of Life and the Origin of the Eukaryotic Cell

Objectives

At the completion of this exercise, students will be able to:

1 Define and apply key terms associated with the origin of life.
2 Describe the general tenets of the RNA world hypothesis.
3 List evidence from current RNA functions that are considered "ghosts of an RNA past."
4 Discuss the origin of organelles and eukaryotic life.
5 Provide evidence that supports the theory of symbiogenesis.
6 Diagram a stepwise progression of the origin of eukaryotic cells and organelles.

ⓘ Background Information

The RNA world hypothesis argues that RNA polynucleotides were the first self-replicating molecules and precursors to all subsequent life on earth. In support of this hypothesis, researchers point to several characteristics of RNA:

1. RNA has catalytic functions in protein synthesis. It is the relatively small lengths of RNA in the ribosome (rRNA), and not the ribosomal proteins, that catalyze peptide bond formation.

2. RNA can carry all the encoded information needed to carry out all the processes necessary for life (think mRNA, tRNA).

3. The most conserved (slowly evolving) and universally important components of cells are composed of RNA (rRNA).

4. RNA molecules are the most important non-protein compounds that participate in energy conversions in the cell (ATP, NADH, etc.).

Thus, the RNA world hypothesis points to the current functions of RNA in the cell as evolutionary remnants, or "ghosts of an RNA past."

✎ **Activity 1** Key Terms

Materials Required
☐ Biology textbook or access to internet resources

Define the following terms:

Abiogenesis

_____

_____

_____

RNA world

_____

_____

_____

Panspermia

_____

_____

_____

Symbiogenesis (endosymbiotic theory)

_____

_____

Last universal ancestor

_____

_____

"Primordial soup"

_____

_____

Activity 2 RNA World Theory and the Chemosynthetic Origin of Life

Materials Required
☐ Access to internet resources

1 Watch the following video:

https://www.youtube.com/watch?v=VYQQD0KNOis (NOVA youtube page)

_OR_

http://www.pbs.org/wgbh/nova/body/rna-origin-life.html (direct NOVA page at PBS)

2 Which current functions of RNA can be considered "ghosts of an RNA past"?

_____

_____

_____

_____

_____

Activity 3 Major Events in Earth History

Materials Required
❏ Biology textbook or access to internet resources

1 How old is the universe?

_____

_____

2 How old is our solar system?

_____

_____

3 What is "primordial soup"?

_____

_____

4 When did prokaryotes first appear on Earth?

_____

_____

5 When did eukaryotes first appear on Earth?

_____

_____

Activity 4 Symbiogenesis and the Origin of Mitochondria and Chloroplasts

Materials Required
❏ Colored pencils, pens, or markers

1 Using the components in Figure 21.1, draw a stepwise model depicting the origin of mitochondria and plastids. Include in your model the following steps:

a. Prokaryotic cell with cell membrane, nucleoid, and cytoplasm.

b. Formation of membrane-bound nucleus from cell membrane infolding—it becomes a eukaryote.

c. An aerobic bacterium enters/is taken up by the cell (parasite/prey), which avoids digestion and becomes an endosymbiont.

d. A cyanobacterium enters/is taken up by the cell.

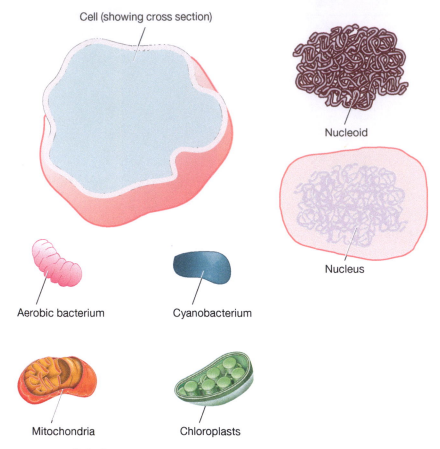

Cell (showing cross section)

Nucleoid

Nucleus

Aerobic bacterium

Cyanobacterium

Mitochondria

Chloroplasts

FIGURE **21.1** ● Components in the origin of mitochondria and plastids.

2 What is the advantage of a nucleus? How could this development give a eukaryotic cell a selective advantage?

3 Which eukaryotic organelle arose from a prokaryotic aerobic bacterium?

4 Why would it be more beneficial for a eukaryote to form a symbiotic relationship with an aerobic bacterium rather than an anaerobe?

5 Which eukaryotic organelle arose from a cyanobacterial endosymbiont?

6 What is the selective advantage of having a cyanobacterial endosymbiont?

7 Using only two complete sentences, describe how Figure 21.2 provides evidence in support of the symbiogenesis theory of eukaryotic evolution.

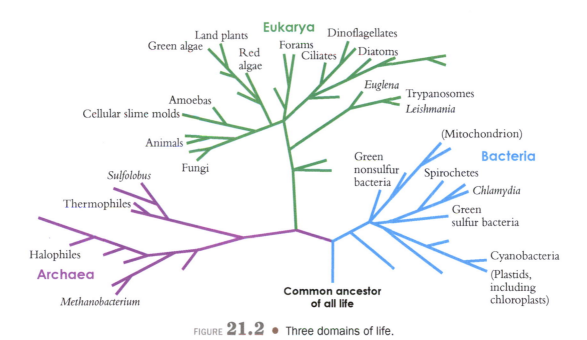

FIGURE **21.2** • Three domains of life.

Activity 5 Synthesis

Materials Required
No materials required for this activity

Now let's tie it all together.

1 Consider the tenets of the RNA world hypothesis. Why don't we see life re-emerging *de novo* on earth now?

2 Consider the tenets of the symbiogenesis theory of the origins of eukaryotic organelles, events that took place as far back as 1.5 billion years ago. If eukaryotic cells have such an advantage over prokaryotic cells, why don't we see the emergence of new organelles now? (***Hint:*** Another way of thinking about the question is to consider why prokaryotic cells don't just independently evolve nuclear membranes and enslave aerobic bacteria and cyanobacteria today.)

3 How do you explain the fact that the DNA of the mitochondria in your cells is more closely related to bacteria than the nuclear DNA of other animals? How do you explain that chloroplast DNA in corn is more similar to cyanobacteria than the nuclear DNA in grass?

4 The RNA world hypothesis is the most widely accepted model of abiogenesis, but it is not the only one. What are some of the other proposed chemosynthetic origins of life? List and provide a brief statement about the supporting ideas of each hypothesis.

Exercise 22
History of Evolutionary Thought

Objectives

At the completion of this exercise, students will be able to:

1 Define and apply key terms and concepts concerning the history of evolutionary thinking.

2 Identify the scientists and publications that influenced Darwin's unifying theory of evolution

3 Describe how the modern theory of evolution reconciles patterns in nature (fossils and the geological record, structural similarity, vestigial traits, and geographic affinities of related species).

4 Distinguish between science and religion, and evaluate the basic claims of "intelligent design."

5 Describe how societal influences played a role in the development of evolutionary thought.

ⓘ Background Information

Despite common perception, most scientific advances are very much shaped by the social context in which they arise. Historically, the prevailing religious, philosophical, and scientific leanings of the day exerted a powerful influence on the types of questions that are asked, how they are answered, and how those answers are interpreted. The history of evolutionary thought is no different, and perhaps provides one of the most compelling examples of the role(s) that the prevailing social climate drives and shapes scientific advances.

Activity 1 Key Terms

Materials Required
❑ Biology textbook or access to internet resources

Define the following terms:

Great chain of being

_____

_____

Systema Naturae

_____

_____

Fixity of species

_____

_____

Adaptation

_____

_____

Inheritance of acquired traits

_____

_____

Extinction

_____

_____

Natural theology

_____

_____

Faunal succession

_____

_____

Uniformitarianism

_____

_____

Sarawak Law

_____

_____

Wallace Line

_____

_____

Developmental homology

_____

_____

Structural homology

_____

_____

Genetic homology

_____

_____

Vestigiality

_____

_____

Atavism

_____

_____

Consilience

_____

_____

Modern evolutionary synthesis

_____

_____

Intelligent design

_____

_____

Irreducible complexity

_____

_____

Specified complexity

_____

_____

Wedge strategy

_____

_____

Activity 2 Darwin's Influences

Materials Required
❑ Biology textbook or access to internet resources

1 Describe how each of the following individuals shaped public perceptions of the natural world:

a. Plato ("essentialism")

_____

_____

_____

b. Aristotle (fixed properties of species)

_____

_____

_____

c. Carl Linnaeus

_____

_____

_____

d. William Paley

_____

_____

_____

e. Robert Chambers

_____

_____

_____

2 Describe how each of the following individuals influenced Darwin's thinking as he developed a unifying theory of biology:

a. Georges Cuvier

b. Charles Lyell

c. Thomas Robert Malthus

d. Richard Owen

e. James Hutton

f. Jean-Baptiste Lamarck

g. George Louis Leclerc, Compte de Buffon

h. Alfred Russell Wallace

3 Describe how each of the following observations influenced Darwin's thinking as he developed a unifying theory of biology:

    a. His discovery of a giant fossil ground sloth in South America. Think about the connection between extinct and extant South American animals.

    _____

    _____

    _____

    b. His observation of tortoises on different islands of the Galapagos having distinct shell shapes, and that birds on the Galapagos Islands were different from but most similar to those in Chile.

    _____

    _____

    _____

    _____

4 Describe how each of the following individuals influenced the perception of Darwin's ideas by the general public:

    a. Michael Behe

    _____

    _____

    _____

    b. William Dembski

    _____

    _____

    _____

    c. U.S. District Judge John E. Jones III

    _____

    _____

    _____

    _____

Activity 3 Patterns and Process

Materials Required
No materials required for this activity

In this activity, you will practice consilience, or reconciling multiple lines of evidence.

1 What pattern is depicted in Figure 22.1?

_____

2 How does the theory of evolution reconcile this pattern?

_____

Era	Period	Age (mya)	Important Events	Representative Organisms
Cenozoic	Quaternary	2.5	Ice ages; extinction of many large mammals; origin of genus *Homo*	
Cenozoic	Tertiary		First primates appear; rapid diversification of mammals, birds, and pollinating insects; flowering plants continue dominance; most present-day mammal orders evolved; Europe separates from North America; primates diversify; first apes; grazing mammals appear as grasses replace forests in drier areas; first apes; first bipedal hominids	
Mesozoic	Cretaceous	65 / 146	Flowering plants diversify and dominate; extinction of dinosaurs	
Mesozoic	Jurassic	200	Dinosaurs at their peak; first bird fossils; first flowering plants	
Mesozoic	Triassic	250	First dinosaurs and mammals; coniferous forests dominate landscape	
Paleozoic	Permian	299	Diversification of reptiles; first mammal-like reptiles; origin of most insect groups	
Paleozoic	Carboniferous	360	First reptiles; first seed plants; large coniferous forests	
Paleozoic	Devonian	416	Diversification of bony fishes; first amphibians and insects appear; first forests	
Paleozoic	Silurian	444	First jawed fishes; invertebrates colonize land	
Paleozoic	Ordovician	500	Radiation of fishes; first land plants and fungi appear	
Paleozoic	Cambrian	570	Rapid diversification of invertebrates into major phyla; first vertebrates (Cambrian explosion)	
Precambrian		4,600	Formation of Earth; oldest known rocks on Earth's surface; first prokaryotes appear; accumulation of atmospheric oxygen; first eukaryotic cells appear; first evidence of multicellular life; soft-bodied invertebrates appear	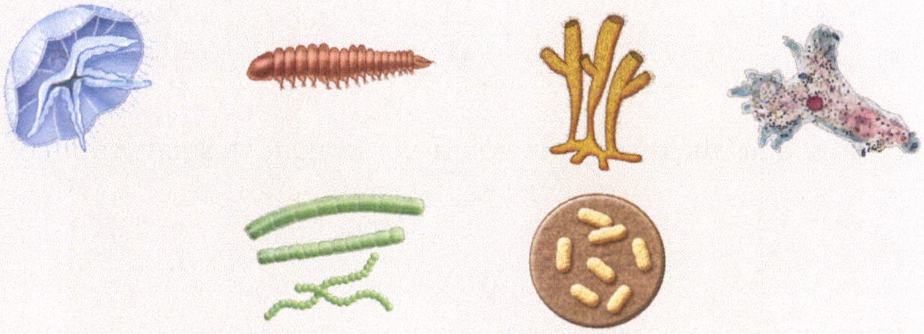

FIGURE **22.1** • Fossil record.

After Smith, D. *Exploring Zoology: A Laboratory Guide*, 2e. Englewood, CO: Morton Publishing Company, 2014

3 What pattern is depicted in Figure 22.2?

4 How does the theory of evolution reconcile this pattern?

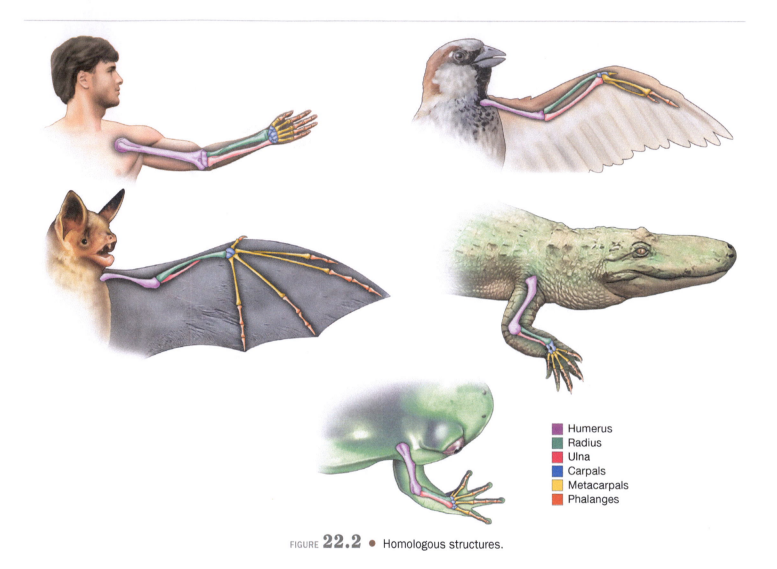

Humerus
Radius
Ulna
Carpals
Metacarpals
Phalanges

FIGURE **22.2** ● Homologous structures.

5 What pattern is depicted in Figure 22.3?

6 How does the theory of evolution reconcile this pattern?

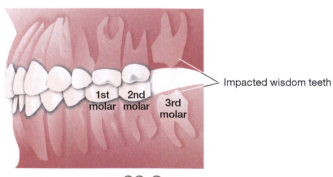

Impacted wisdom teeth

1st molar 2nd molar 3rd molar

FIGURE **22.3** ● Molars.

7 What pattern is depicted in Figure 22.4? Think about Wallace's Line.

8 How does the theory of evolution reconcile this pattern?

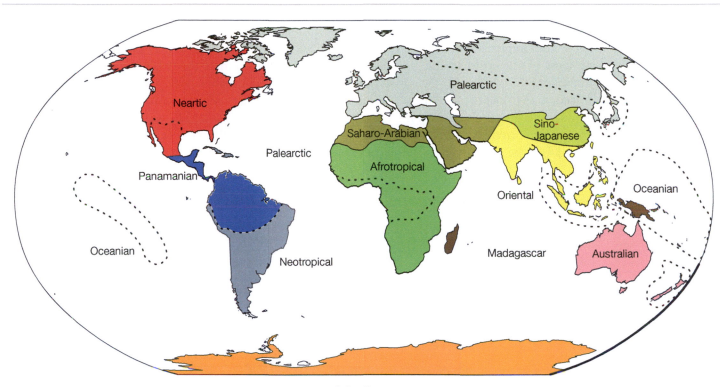

FIGURE **22.4** ● World map.

Materials Required

❑ Colored pencils, pens, or markers

Science and religion are just two among many ways of knowing. Stephen Jay Gould, a prominent evolutionary biologist, characterized the difference between science and religion in the following way: "Science tries to document the factual character of the natural world, and to develop theories that coordinate and explain these facts. Religion, on the other hand, operates in the equally important, but utterly different, realm of human purposes, meanings, and values—subjects that the factual domain of science might illuminate, but can never resolve."[1] For some, religion may be an important way of processing and making sense of the world around us, but it is important not to conflate the two enterprises. Indeed, in 1999, the National Academy of Sciences noted, "Scientists, like many others, are touched with awe at the order and complexity of nature. Indeed, many scientists are deeply religious. But science and religion occupy two separate realms of human experience. Demanding that they be combined detracts from the glory of each."[2] Proponents of intelligent design claim that intelligent design provides a scientific alternative to evolution. Is intelligent design a scientific enterprise, or is it a religious enterprise? Use the prompts that follow to explore answers to this question.

[1] Gould, S. J. (1999). *Rocks of Ages: Science and Religion in the Fullness of Life* (1st ed.). New York: Ballantine Pub. Group.

[2] National Academy of Sciences (U.S.). (1999). *Science and Creationism: A View from the National Academy of Sciences* (2nd ed.). Washington, DC: National Academy Press.

1 Scientific arguments must focus on observations of the natural world and cannot employ supernatural explanations. Does intelligent design invoke the supernatural?

_____

_____

2 Scientific theories must be subject to falsification. What evidence or observations would be sufficient or necessary for you to reject the "hypothesis" of intelligent design?

_____

_____

_____

3 Scientific theories are predictive. For example, if the preponderance of evidence suggests that humans originated in Africa over 200,000 years ago, then we would predict that we shouldn't find any fossil human skeletons in Utah prior to that time. What testable predictions about the biological world are made by intelligent design?

_____

_____

4 Another characteristic of science is that it builds on previous knowledge and produces new knowledge.

a. What prior knowledge is intelligent design based on?

_____

b. What new theories or disciplines have emerged from creation research?

_____

5 How do proponents of intelligent design use a mousetrap as an example of irreducible complexity?

_____

_____

_____

_____

_____

_____

6 Draw a picture that illustrates how a mousetrap, missing one of its parts, can still serve a useful purpose.

Activity 5 Synthesis

Materials Required
No materials required for this activity

Now let's tie it all together.

1 What do we mean by "descent with modification"?

_____

_____

_____

2 What's the difference between an atavism and a vestige? Provide an example of each.

_____

_____

_____

3 Why were Cuvier's ideas about extinction controversial?

4 Why did Robert Chambers publish his book, *Vestiges of the Natural History of Creation,* anonymously?

5 Why did Darwin believe that Chambers' book prepared the public to more favorably receive his book, *On the Origin of Species by Means of Natural Selection, or the Preservation of Favoured Races in the Struggle for Life?*

6 If Alfred Russell Wallace independently discovered and described natural selection as a mechanism of evolution, why does Darwin usually get most of the credit?

7 What are the similarities between intelligent design and natural philosophy? How are they different?

Exercise 23
Tree Thinking: Phylogenetics

Objectives

At the completion of this exercise, students will be able to:

1 Define and apply phylogenetic terminology.

2 Discuss cladistic and phylogenetic terms and their relevance/importance to reconstructing evolutionary history.

3 Build a rudimentary phylogenetic tree from a given transformation series.

4 Use concepts of homology and analogy to infer monophyly and homoplasy.

5 Map the origin and evolution of characters on a phylogenetic tree.

6 Read, manipulate, and analyze an evolutionary tree.

ⓘ Background Information

One of Charles Darwin's most significant contributions to evolutionary theory is "tree thinking," so named because Darwin considered a tree to be an apt metaphor for the evolution of life on Earth. As with a tree, Darwin's idea of evolution (descent with modification) is hierarchical, yet always growing. Portions of the tree expand and flourish while other portions die off, and all the branches of the tree are derived from a single common ancestor (the trunk of the tree).

Because we rarely have direct measures of evolutionary history, scientists interested in studying the evolutionary relationships among organisms, or the origin and maintenance of a specific trait, must reconstruct evolutionary history. A figure that depicts the evolutionary relationships among species (or higher taxa) is called a phylogenetic tree.

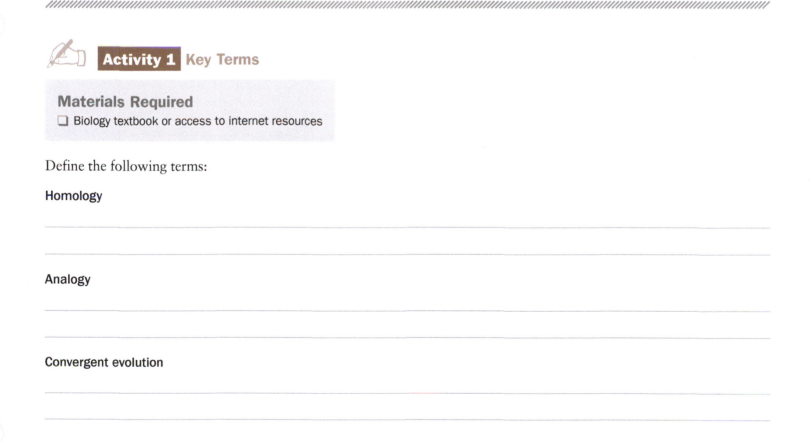

Activity 1 Key Terms

Materials Required
☐ Biology textbook or access to internet resources

Define the following terms:

Homology

_____

_____

Analogy

_____

_____

Convergent evolution

_____

_____

_____

Monophyletic group

Paraphyletic group

Polyphyletic group

Taxon

Homologous character

Apomorphy

Plesiomorphy

Non-homologous (analogous) character

Homoplasy

Ingroup

Outgroup

Sister group

_____

_____

Character polarity

_____

_____

_____

Clade

_____

_____

_____

Cladogram

_____

_____

_____

Parsimony

_____

_____

_____

Most recent common ancestor (MRCA)

_____

_____

_____

Polytomy

_____

_____

_____

Activity 2 Building a Phylogenetic Tree

Materials Required
❑ Blue, red, and black pens or markers

Table 23.1 is a data matrix that summarizes some of the key traits of the chordate Superclass Tetrapoda, which includes the four-limbed vertebrates and their descendants, including the living and extinct amphibians, reptiles, birds, and mammals. In the character matrix, the characters in each column are assumed to be homologous. For example, the hair/fur of dogs and humans is similar because it was present in their shared/common ancestor. A zero (0) indicates that the character is absent, and a one (1) indicates that it is present.

    Note that the dog and the human share a unique, derived character (unique = apomorphy; shared = synapomorphy) to the exclusion of the other taxa: both lactate and have fur. We know that lactation and fur are derived, and not ancestral character states, because they differ from the outgroup taxon, which is assumed to always possess the ancestral state. Thus, compared against all the other taxa in the matrix, dogs and humans share a most recent common ancestor and are each other's nearest relatives. Synapomorphies are used to reconstruct evolutionary history.

1 Use the information in Table 23.1 to reconstruct the evolutionary history of these organisms (taxa).

TABLE **23.1** Data Matrix for Various Chordates

Taxon	Fur, lactation	4 limbs	Amniotic egg; fingernails on digits	Gas exchange across skin	Egg with mineralized outer membrane
Human	1	1	1	0	0
Dog	1	1	1	0	0
Chicken	0	1	1	0	1
Lizard	0	1	1	0	1
Frog (outgroup)	0	1	0	1	0

2 There are many different ways to reconstruct phylogenetic trees, but the following procedure is one example:

a. In the space provided in Figure 23.1, begin building your tree. We will help you get started by writing "human" and "dog" next to each other. Below them, we will draw a blue dot to represent their common ancestor, and then draw a line between the ancestor and each descendent. We will use a red line or tick mark to designate the synapomorphy that unites the two taxa.

b. Note that the chicken and the lizard also share an apomorphic character (egg with mineralized outer membrane). Draw a blue circle in sketch box B to represent the common ancestor of the chicken and the lizard, and link them together with black lines. Designate their synapomorphy with a red tick mark (red line).

c. Next, look at the characters "amniotic egg; fingernails (horny nails on digits)." Note that this is a synapomorphy that unites the human, dog, chicken, and lizard to the exclusion of the frog. In sketch box C, depict this relationship the same way you did in step b (you'll have four black lines connecting your taxa to their most recent common ancestor and a single red tick mark indicating their synapomorphy).

d. Next, look at the character "4 limbs." In sketch box C, draw the relationship that is supported by this synapomorphy.

e. Notice that the trees for fur and lactation, amniotic egg, and egg with mineralized outer membrane, are combinable components that do not contradict, but instead further clarify or resolve the relationships established by the character "4 limbs." Combine each of the trees (components) into a single tree in sketch box C. Be sure to indicate the synapomorphies that support the position of each shared branch in the tree.

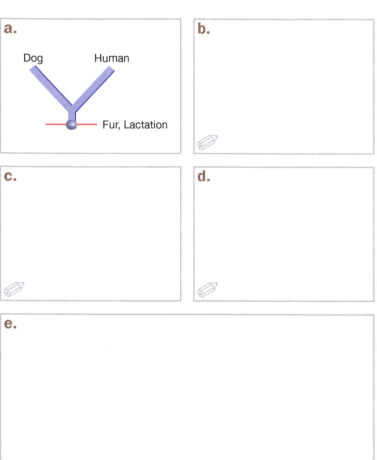

FIGURE **23.1** ● Synapomorphy.

3 Answer the following questions:

a. Which group is most closely related to the dog and human? That is, which groups share a MRCA with mammals? Explain.

b. According to your tree, are mammals "more highly evolved" than lizards and birds? Frogs? Why or why not?

c. Snakes are more closely related to lizards than birds, but they don't have legs. Aren't they more appropriately placed in a clade that includes earthworms and other legless organisms? Why or why not?

d. Frog eggs don't have an amnion; their eggs have only a single membrane and must be laid in water. Is it correct to refer to frog eggs as having the plesiomorphic character state, in comparison to amniotes?

e. If frog eggs are plesiomorphic, is it correct to refer to extant frogs as "ancestral" or "primitive" relative to amniotes? Explain.

Materials Required
❑ Blue, red, and black pens or markers

The tree in Figure 23.2 depicts the current best estimate of the evolutionary relationships among the tetrapods and includes "fish" (which don't have limbs) as the outgroup taxon.

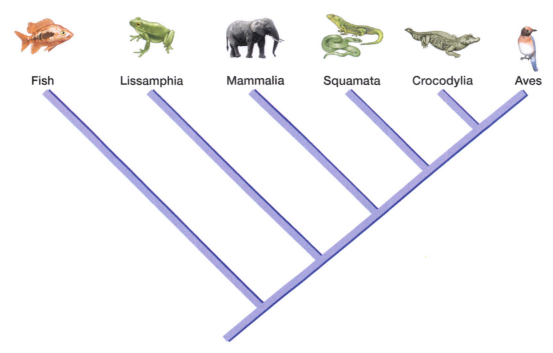

FIGURE **23.2** ● Phylogenetic relationships among the Tetrapoda.

1 Add a red tick mark to the tree in Figure 23.2, and label it "tetrapod limbs" at the point where four limbs evolved in an ancestral population.

2 Mammals, lizards, snakes, turtles, crocodiles, and birds all have an amnion, which is an embryonic membrane that transports oxygen into the egg and expels carbon dioxide. Add a red tick mark on the tree in Figure 23.2, and label it "amniotic egg" to indicate the position on the tree where the amniotic membrane evolved in tetrapods.

3 Answer the following questions:

a. Which group is most closely related to mammals: frogs or birds? Explain. (**Hint:** Consider which group shares a most recent common ancestor with mammals.)

_____

_____

_____

_____

_____

_____

b. Which group is most closely related to turtles? Explain.

_____

_____

_____

_____

_____

c. In the tree in Figure 23.2, is the character "lactation" a synapomorphy? Explain your reasoning.

_____

_____

_____

_____

_____

_____

4 In the space provided, redraw the tree exactly as it appears in Figure 23.2, but swap the positions of the crocodiles and birds. Does this swap in positions change the way we interpret their phylogenetic relationship? Now swap the positions of the lizards and crocodiles. Does this swap in positions change the way we interpret their phylogenetic relationships? Justify your reasoning.

 **Activity 4** Synthesis

Materials Required
No materials required for this activity

Now let's tie it all together.

1 Identify which of the following pairs of similar-looking traits are homologous and which are analogous:

a. The forelimbs of a penguin and a chicken.

_____

_____

b. The streamlined, torpedo shape of sharks and whales.

_____

_____

c. The eye position on the skull of crocodiles and hippopotami.

_____

_____

d. The camera-style eye of an octopus and the lens-shaping eye of a human.

_____

_____

e. The forelimbs of a pterodactyl and a bird.

_____

_____

f. The wings of bats and birds.

_____

_____

g. The humerus of a horse and a sea lion.

_____

_____

2 How could the use of analogous traits versus homologous traits influence phylogenetic reconstruction of evolutionary history?

_____

_____

_____

_____

3 Figure 23.3 is a phylogenetic tree constructed from the data matrix in Table 23.2. Use the information in the figure and the data matrix to answer the questions that follow. Note that a zero (0) indicates that the character is absent, and a one (1) indicates that the character is present.

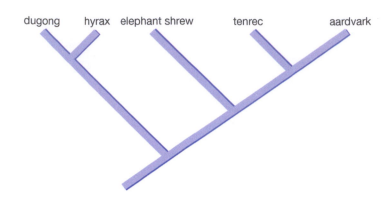

FIGURE **23.3** • Phylogenetic tree based on data matrix in Table 23.2.

TABLE **23.2** Data Matrix (Transformation Series)

	Characters (numbered)				
	1	2	3	4	5
Dugong	1	0	0	1	1
Hyrax	1	0	0	1	1
Aardvark	1	1	1	0	0
Tenrec	1	1	1	0	0
Elephant shrew	1	1	0	0	0
Elephant	1	0	0	1	0

a. Based on the data in Table 23.2, add elephants to the tree.

b. Which is the sister taxon/group to elephants?

c. Which character supports the hypothesis that all of the taxa in the tree form a clade (monophyletic group)?

d. Tenrecs and aardvarks are exclusively insectivorous. Map the origin of insectivory on the tree.

e. Whales and dugongs have streamlined bodies, flippers, paddle-like tails, and very small amounts of fur. Based on the data in the transformation series, are the similarities between whales and dugongs due to homology or homoplasy?

f. According to the tree, which taxon/taxa represent the nearest living relative to elephant shrews?

Enhancement Exercises for Biology • **UNIT 6:** Evolution and Origin of Species

Exercise 24
Mechanisms of Evolution

Objectives

At the completion of this exercise, students will be able to:

1 Define five mechanisms of evolution, and describe how they can work together or in opposition to each other.

2 Distinguish the levels at which natural selection acts and evolution occurs.

3 Describe the utility of the Hardy-Weinberg equilibrium model for studying evolution.

4 Discuss the assumptions of the Hardy-Weinberg equilibrium model and the postulates of adaptive evolution.

5 Address common misconceptions about evolution.

6 Identify modes of natural selection.

ⓘ Background Information

Mutation, migration, genetic drift, and non-random mating are all mechanisms of evolution. Each can play a role in generating and maintaining genetic variation in populations, but the strongest driver of evolutionary change is natural selection. (Remember that evolutionary change is defined as a change in allele frequency of time.)

In order for adaptive evolution to occur, four postulates of natural selection must be met:

If...

1. Variation exists within populations (among individuals),

2. Variation is heritable (at least some of it),

3. Each generation has differential survival and/or reproduction (among individuals),

4. Survival and reproduction are not random (some individuals have favorable variations that lead them to survive and/or reproduce better),

Then... adaptive evolution can occur.

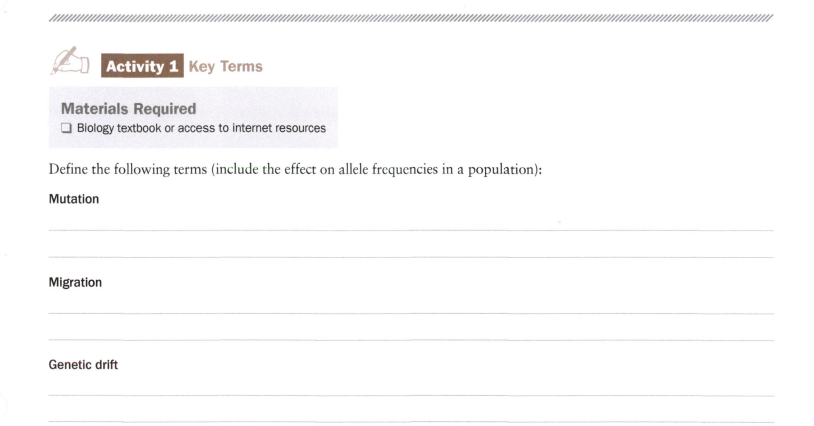

Activity 1 Key Terms

Materials Required
❏ Biology textbook or access to internet resources

Define the following terms (include the effect on allele frequencies in a population):

Mutation

_____

_____

Migration

_____

_____

Genetic drift

_____

_____

Non-random mating

_____

_____

Natural selection

_____

_____

Activity 2 The Hardy-Weinberg Equilibrium Model

Materials Required
☐ Biology textbook or access to internet resources

1 Write out the equation for the Hardy-Weinberg equilibrium model.

_____

2 What are the assumptions of the Hardy-Weinberg equilibrium model?

_____

_____

_____

3 What good is the Hardy-Weinberg equilibrium model to an evolutionary biologist?

_____

_____

_____

_____

Activity 3 Natural Selection on Bird Beak Length

Materials Required
☐ 4 wooden craft sticks
☐ 2 rubber bands
☐ 10 pennies
☐ Timer
☐ 25 large grapes
☐ 25 small raisins
☐ 1 plate
☐ 2 cereal bowls

While in the Galapagos Islands, Darwin noticed that the beaks of various finches were different lengths, shapes, and sizes. The differences he observed led him to several of his most important ideas about the mechanisms and processes that are responsible for descent with modification, including natural selection.

Sialia mexicana (the Western Bluebird) is a small passerine bird in the family Turdidae. It has a thin, straight beak and feeds primarily on worms, berries, and insects, but it is also known to be a voracious consumer of grapes and raisins. In this exercise you will simulate the functional consequences of variation in beak shape (short, stout beaks to long, thin beaks) of the Western Bluebird and how these might be reflected in subsequent generations. For the purposes of this activity, let's assume:

- ☐ There is variation in beak length among individuals in a population.
- ☐ Variation in beak length is heritable.
- ☐ Individuals with one type of beak length survive or reproduce better than others.

1 Make your short, stout beak (Fig. 24.1) by placing 3 pennies between two of the wooden craft sticks about 1 inch (2.54 cm) from one end.

2 Secure the wooden craft sticks together with one of the rubber bands.

3 Now make your long, thin beak (Fig. 24.2) by placing 7 pennies between the remaining two wooden craft sticks, this time approximately 2½ inches (6.35 cm) from one end, and secure with the final rubber band.

4 Spread the raisins evenly over the surface of the plate, and set your timer for 30 seconds.

5 For 30 seconds, attempt to pick up as many raisins as possible with your small beak and place them in the cereal bowl.

6 Record the number of raisins picked up with the small beak: _____

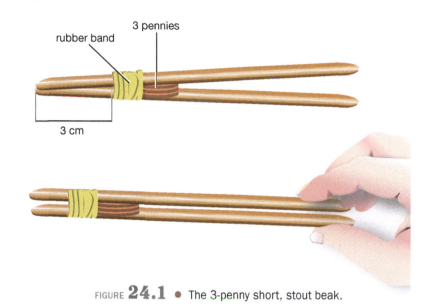

FIGURE **24.1** ● The 3-penny short, stout beak.

7 Return the raisins to the plate.

8 Now attempt to pick up as many raisins as possible with the large beak and place them in the cereal bowl in the same amount of time.

9 Record the number of raisins picked up with the large beak: _____

10 Clear the plate.

11 Spread the grapes out evenly on the plate, and again, set your timer for 30 seconds.

12 Using your small beak, attempt to pick up as many of the grapes as possible in 30 seconds and place them in the bowl.

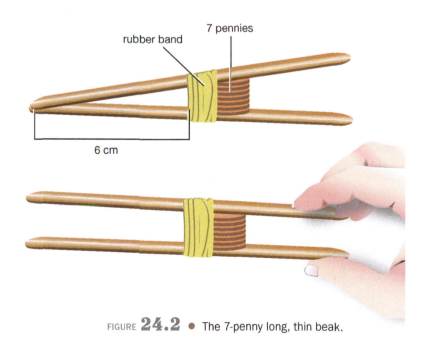

FIGURE **24.2** ● The 7-penny long, thin beak.

13 Record the number of grapes picked up with the small beak: _____

14 Now, using the large beak, attempt to place as many of the grapes as possible in the bowl in the same amount of time.

15 Record the number of grapes picked up with the large beak: _____

16 Return the grapes to the plate.

17 In the space provided, generate a graph of your data that allows for a comparison of the efficiency of the two beak types at collecting grapes and raisins.

18 Assuming that the grapes represent the most common food source available to the Western Bluebird to be eaten in a normal year, which beak is better suited for eating/retrieving the grapes?

_____

19 Now imagine that a severe drought results in there being numerous raisins available, but very few grapes. Which beak is better suited for eating/retrieving raisins?

_____

20 Following a drought year, how would you predict the frequency/abundance of long-beaked Western Bluebirds to change?

_____

_____

21 Does the predicted change in abundance of birds with short, stout beaks reflect stabilizing, disruptive, or directional selection? (*Hint:* Think about modes of selection.)

_____

22 If grapes are the most common food source for most of the years, and drought occurs infrequently, how could there be variation in beak length?

_____

_____

_____

_____

Activity 4 Synthesis

Materials Required
No materials required for this activity

Now let's tie it all together.

1 Does selection act on individuals or populations? Do individuals evolve, or do populations? Explain your answers using the Western Bluebird beak length selection activity above as an example.

_____

_____

_____

_____

2 Describe a situation in which migration and natural selection could work in opposition to each other. Think about how the net change in allele frequencies is zero.

_____

_____

_____

_____

3 Consider the four postulates of natural selection. How have humans manipulated each of these postulates (in natural or artificial systems)?

_____

_____

_____

_____

4 Make a list of human traits that are NOT adaptive (not under selection).

_____

_____

_____

_____

5 What is wrong with the statement "weak elk give themselves up to wolves for the good of the species"? (**Hint:** Is natural selection teleological?)

_____

_____

_____

_____

6 Most mutations (source of new alleles) are random events. Why is it incorrect to say that natural selection, or adaptation, is random?

_____

_____

_____

_____

Exercise 25
Species Concepts and Speciation

Objectives

At the completion of this exercise, students will be able to:

1 Define and apply key terms associated with species concepts and speciation.

2 Define and argue the strengths and weaknesses of five prominent species concepts.

3 Calculate changes in allele frequencies due to genetic drift as a result of allopatric vicariance.

4 Make predictions about how allele frequencies will respond when populations are reproductively isolated.

5 Understand how pre- and post-zygotic mating barriers (reinforcement) have evolved to avoid the deleterious consequences of hybridization.

6 Use a phylogeny to test predictions about the mode and relative tempo of speciation.

ⓘ Background Information

The "species problem" refers to the persistent challenge biologists face when trying to define what species are, and how to delimit them. This is due primarily to the fact that it is very difficult to simultaneously define what a species is (ultimate reality, or ontology) and how to delimit them (discovery operations, or epistemology). Accordingly, scientists struggle to agree on a single concept of species that is both philosophically satisfying and at the same time "works" when trying to apply it across all living (or formerly living) things.

Activity 1 Key Terms

Materials Required
☐ Biology textbook or access to internet resources

Define the following terms:

Biological species concept

_____

_____

_____

_____

_____

Typological (morphological/Linnean) species concept

Phylogenetic species concept

Ecological species concept

Evolutionary species concept

Speciation

Vicariance

Cryptic species

_____

_____

Gene flow

_____

_____

Hybridization

_____

_____

Hybrid zone

_____

_____

Prezygotic mating barrier

_____

_____

Postzygotic mating barrier

_____

_____

Allopatric speciation

_____

_____

Peripatric speciation

_____

_____

Parapatric speciation

_____

_____

Sympatric speciation

_____

_____

Activity 2 Species Concepts

Materials Required
No materials required for this activity

1 List one strength and one weakness of each species concept. Think about potential problems or sources of error with the definition or the means by which they are delimited.

    a. Typological Species Concept (strength)

_____

_____

    Typological Species Concept (weakness)

_____

_____

    b. Biological Species Concept (strength)

_____

_____

    Biological Species Concept (weakness)

_____

_____

    c. Phylogenetic Species Concept (strength)

_____

_____

    Phylogenetic Species Concept (weakness)

_____

_____

    d. Ecological Species Concept (strength)

_____

_____

    Ecological Species Concept (weakness)

_____

_____

e. Evolutionary Species Concept (strength)

_____

_____

_____

Evolutionary Species Concept (weakness)

_____

_____

_____

_____

Activity 3 Speciation

Materials Required

- ❏ 3 plastic cups (1 large one and 2 smaller ones)
- ❏ 1 small bowl
- ❏ 50 white beans*
- ❏ 50 red beans*

* Can substitute small candies, such as Skittles, M&M's, gumdrops, etc., so long as you have at least 50 each of 2 different colors

1 Place all of the red and white beans into a single large cup. The red and white beans in the large cup represent the distribution of two alleles of a single-locus gene in a single, large, parental population of Antarctic springtails (small, six-legged, free-living, omnivorous, and really tough arthropods). Each red bean will represent a big R allele, which is dominant, and each white bean will represent a little r allele, which is recessive. Table 25.1 shows the initial allele frequencies in terms of their genotype and phenotypic frequencies in the parental population.

TABLE **25.1** Initial Allele Frequencies		Original Population	
Genotype	**Number of each individual genotype**	**Genotype (percent)**	**Phenotype (percent)**
YY	25	25	75
Yy	50	50	
yy	25	25	25

2 Imagine that a large glacier crosses through the middle of the parental population such that the population on one side of the glacier can no longer mate with individuals in the population on the other side of the glacier. The two smaller cups represent the descendent populations, one on each side of the glacier.

3 Label one of the smaller cups "North" to represent the population that is now on the north side of the glacier. Randomly draw 50 beans from the parental cup, and place them in the cup labeled "North."

4 Label the second small cup "South" to represent the population that is now on the south side of the glacier. Place the remaining 50 beans in the second small cup labeled "South."

5 Next, keeping the two populations separate, allow individual springtails in each of the populations to mate and produce offspring. To do this, close your eyes and randomly select two beans from the cup labeled "North" and place them in the small bowl. The two beans in the bowl represent the paternal and maternal alleles in their offspring (their newly hatched chick).

6 Open your eyes and tally the genotype of each mating in Table 25.2. Perform 10 matings.

TABLE **25.2** "North" Population

Genotype	Number of each individual genotype	"North" Population	
		Genotype (percent)	Phenotype (percent)
RR			
Rr			
rr			

7 Repeat steps 3 and 4, but this time mating and drawing "alleles" from the cup labeled "South." Keep track of the results of each mating in Table 25.3.

TABLE **25.3** "South" Population

Genotype	Number of each individual genotype	"South" Population	
		Genotype (percent)	Phenotype (percent)
RR			
Rr			
rr			

8 Although the red beans and the white beans (alleles) shared the same frequency in the parental population, due to random chance alone, it is highly likely that some alleles on the north side of the glacier are now at a different frequency than they were in the parental population. If so, you just witnessed a change in allele frequency over time. Evolution just happened!

9 Now imagine that big R and little r are alleles associated with the springtail's ability to produce antifreeze proteins—small proteins that bind to water molecules as they start to freeze, which prevent the water molecules from binding to other water molecules and forming ice crystals inside their cells. Let's assume that the big R allele codes for a form of antifreeze protein that is very costly to produce, but also prevents ice crystals from forming at lower temperatures than the antifreeze proteins produced by the little r allele.

10 If temperatures are usually much colder on the southern side of the glacier, how might natural selection alter allele frequencies in the southern population that were initially shaped only by genetic drift? Explain.

11 In the same scenario, how might natural selection alter allele frequencies in the northern population?

_____

_____

_____

_____

_____

12 At what point is there evidence that the two populations have diverged sufficiently to delimit them as separate species according to the following species concepts:

a. Phylogenetic species concept

_____

_____

_____

_____

b. Biological species concept

_____

_____

_____

_____

c. Ecological species concept

_____

_____

_____

_____

d. Typological species concept

_____

_____

_____

_____

_____

e. Evolutionary species concept

_____

_____

_____

_____

_____

Activity 4 Reproductive Isolation and Reinforcement

Materials Required
No materials required for this activity

When reproductively isolated populations come in contact again, there are four possible mating outcomes:
1. They cannot reproduce and do not hybridize.
2. They are able to hybridize, and it is "bad" (low fitness relative to the parents).
3. They are able to hybridize, and it is "good" (improved fitness relative to the parents).
4. They are able to hybridize and have roughly equal fitness with the parents.

1 Under which of these four scenarios is a prezygotic mating barrier most likely to arise? Explain.

_____

_____

_____

_____

_____

2 Provide an example of a prezygotic mating barrier, and explain how it works to reinforce reproductive isolation in an area of secondary contact (hybrid zone).

_____

_____

_____

_____

_____

_____

3 Provide an example of a postzygotic mating barrier, and explain how it works to reinforce reproductive isolation in an area of secondary contact.

_____

_____

_____

_____

_____

_____

Activity 5 Synthesis

Materials Required

No materials required for this activity

Now let's tie it all together.

1 Apply the following situations to the five main concepts of species. Describe their strengths or weaknesses in terms of their ability to account for:

a. Isolated populations that never come in contact (in nature)

_____

_____

_____

_____

_____

_____

b. Asexual species

_____

_____

_____

_____

_____

_____

c. Hybrid offspring

_____
_____
_____
_____
_____

d. Fossils

_____
_____
_____
_____
_____

e. Operational practicality/applicability

_____
_____
_____
_____
_____

2 Why does speciation require reproductive isolation? What is the effect of reproductive isolation on allele frequencies?

_____
_____
_____
_____
_____

3 How is speciation by dispersal different from speciation by vicariance? How are they similar?

_____
_____
_____
_____
_____

4 The diagram in Figure 25.1 represents the geographic range of three Antarctic tardigrade species (A, B, and C), which are separated by a mountain range and a massive glacier. You hypothesize that these species descended from a common ancestor via dispersal. Specifically, you propose that the original species was confined to the eastern side of the glacier. The first dispersal event brought individuals across the northern part of the glacier. The glacier and the mountain range prevented them from dispersing farther south. Sometime later, a second dispersal event brought individuals across the southern end of the glacier and into the southern mountains. Dispersal of these individuals was also stopped by the glacier and the mountain range to the north. To test your hypothesis, you perform a phylogenetic analysis. In the space provided, draw the phylogenetic tree that supports your hypothesis.

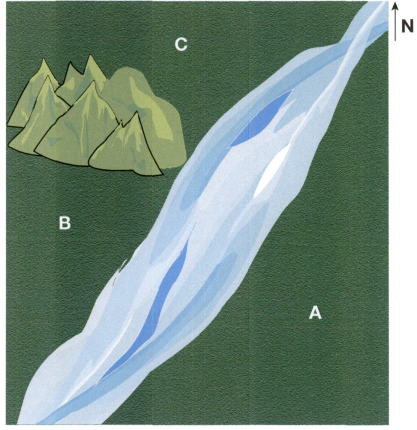

FIGURE **25.1** ● Geographic range of 3 Antarctic tardigrade species.

Diversity of Nature

Exercise 26
The Tree of Life: Survey of Bacteria and Archaea

Objectives

At the completion of this exercise, students will be able to:

1 Define and apply key terms associated with the tree of life.

2 Reconstruct the major branches of the tree of life.

3 Identify the origins of mitochondria and chloroplasts.

4 Compare and contrast the major similarities and differences between bacterial, archaeal, and eukaryotic cells.

5 Recognize the importance of archaea and bacteria in Earth's history and contemporary ecosystem structure and functioning.

6 Characterize any organism in the tree of life by its source of energy and carbon.

ⓘ Background Information

Early attempts to classify all life on earth included the five- or six-Kingdom hypotheses, until Carl Woese argued that phylogenetic relationships based on DNA sequences of the small ribosomal subunit (16s rRNA gene in prokaryotes; 18s rRNA gene in eukaryotes) were inconsistent with these arrangements. Because the phylogenetic tree depicted archaea, a prokaryote lineage, as being more closely related to eukaryotes than to other prokaryotes (bacteria), any classification system that lumped the two prokaryote groups together would be inconsistent with evolutionary history— a phylogenetic lie! To correct this, Woese introduced the three-domain system. Rather than doing away with the previous kingdom system of classification, the three-domain system, which has been widely accepted since the early 1990s, simply provides a level of classification above the kingdom level that is consistent with our best estimate of evolutionary history.

Activity 1 Key Terms

Materials Required
❏ Biology textbook or access to internet resources

Define the following terms:

70s ribosome

_____

_____

80s ribosome

_____

_____

Horizontal gene transfer

_____

_____

Symbiogenesis

Activity 2 **Attributes of Bacteria, Archaea, and Eukarya**

Materials Required
❑ Biology textbook or access to internet resources

1 Using your textbook or internet resources, complete Table 26.1, which summarizes the similarities and differences among the three domains.

TABLE **26.1** Possible Methods for Organisms to Make ATP

	Phototroph: Gets energy from sunlight	**Chemoorganotroph:** Gets energy from organic molecules	**Chemolithotroph:** Gets energy from inorganic molecules (i.e., rocks)
Autotroph: Complex organic compounds are self-synthesized from CO_2, CH_4, and other simple carbon compounds	**Photoautotroph:** Uses sunlight to make its own food from simple carbon compounds		
Heterotroph: Complex organic compounds are synthesized from organic molecules produced by other organisms			

Activity 3 **The Tree of Life**

Materials Required
❑ Colored pens, pencils, or markers
❑ Biology textbook or access to internet resources

The tree of life is a metaphor for depicting the relationships among all living things, from a last universal common ancestor (LUCA) to the present. Championed by Charles Darwin, the tree metaphor is still widely regarded as the best way to depict phylogenetics and evolutionary history. As our picture of the diversity of life on Earth continues to develop, it also becomes more apparent that the vast majority of Earth's diversity is microbial and that plants and animals comprise only a tiny fraction of the Earth's evolutionary and functional diversity.

1 Using internet resources or your textbook, use the terms listed below to fill in the missing information on Figure 26.1. You may wish to color-code them and further annotate the figure as a study aid.

☐ Progenote (LUCA)   ☐ Non-photosynthetic protists   ☐ Protista
☐ Ancestral eukaryotic cell   ☐ Archaea   ☐ Plantae
☐ Mitochondria   ☐ Animalia   ☐ Bacteria
☐ Chloroplasts   ☐ Fungi

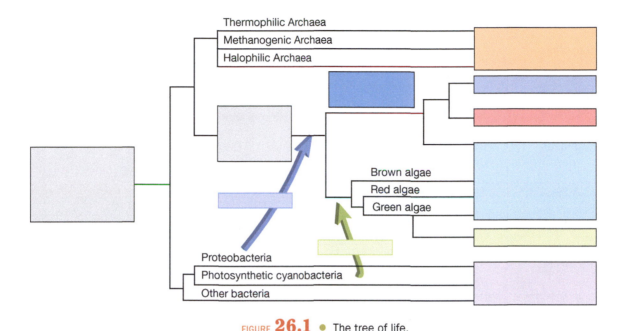

FIGURE **26.1** ● The tree of life.

Activity 4 Functional Diversity of Prokaryotes

Materials Required
❏ Colored pens, pencils, or markers
❏ Biology textbook or access to internet resources

Perhaps the most important and astonishing aspect of bacteria and archaea is the diversity of ways they acquire energy. As a group, they can make their own food via photosynthesis or pull energy out of organic and even inorganic compounds. Of the six ways of synthesizing ATP, eukaryotes are capable of only two. This diversity allows bacteria and archaea to inhabit a much wider array of Earth's habitats than the eukaryotes. The metabolic products of these organisms have shaped the chemistry of earth's oceans, atmosphere, and terrestrial environments. The bacteria and archaea play a critical role in facilitating and regulating the functioning of Earth's ecosystems.

1 Using your textbook or internet resources, complete Table 26.2 to summarize the possible ways that organisms can make ATP. Use your different colored markers to help you keep track of the source of energy used (for synthesis of ATP) and the source of the carbon-carbon bonds, which are needed to synthesize more complex organic compounds. Be sure to briefly define the terms in each box.

TABLE **26.2** Characteristics of Bacteria, Archaea, and Eukarya

Characteristic	Bacteria	Archaea	Eukarya
Nucleus			Present
Organelles		Absent	
Peptidoglycan in cell walls			Absent
Size of ribosomes			
Initiator tRNA			Methionine
RNA polymerases		One*	
Ribosomes sensitive to chloramphenicol and streptomycin			No
Ribosomes sensitive to diptheria toxin	No		

* The archaeal RNA polymerase is more similar to eukaryotic polymerases than bacterial polymerases.

Activity 5 Global Ecosystem Engineering

Materials Required
❑ Biology textbook or access to internet resources

For the first 2.3 billion years, Earth had practically zero available diatomic oxygen (O_2). But then, about 2.7 to 2.5 billion years ago, cyanobacteria began to proliferate in Earth's oceans. By way of photosynthesis, these organisms started cranking out oxygen as a byproduct of fixing carbon. (Recall that $6H_2O + 6CO_2 \rightarrow C_6H_{12}O_6 + 6O_2$). The oxygen concentrations in the oceans and atmosphere ramped up to the point where O_2 was readily available as a final electron acceptor for cellular respiration. This paved the way for aerobic respiration, as opposed to the far less efficient anaerobic respiration, to become possible. Thus, the entire planet was completely altered by a single group of organisms—cyanobacteria.

Archaea and Bacteria continue to play massive roles in regulating the movement of energy and nutrients through Earth's biosphere. One example of this is the critical role that bacteria and archaea play in the nitrogen cycle—the movement of nitrogen atoms through different molecular forms. Nitrogen is an important element of plant fertilizer, but when nitrate runs off of agricultural fields and into groundwater, lakes, rivers, and oceans, it can become a dangerous pollutant. Cyanobacteria thrive in the nitrogen-rich water, and sink to the bottom of the ocean where they become food for the decomposers (other bacteria). The decomposers use up all the oxygen such that the surrounding waters become devoid of oxygen, and all the organisms that require oxygen for aerobic respiration die off.

1 Figure 26.2 depicts the nitrogen cycle. Using your textbook or internet resources, use the following terms to indicate where in the cycle each action takes place:

a. Nitrogen fixation by bacteria and archaea.

b. Decomposition by bacteria and archaea.

c. Oxidation by bacteria and archaea.

d. Oxidation by bacteria.

e. Reduction by bacteria and archaea.

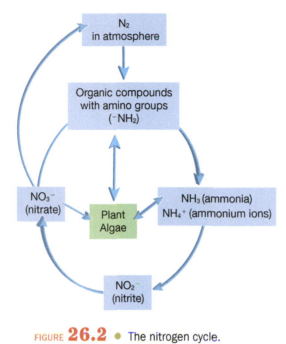

FIGURE **26.2** ● The nitrogen cycle.

2 Annotate the nitrogen cycle to indicate where animals ingest nitrogen (think amino groups/proteins) from plants and where animals release nitrogen (amino groups/proteins or ammonia).

Activity 6 Synthesis

Now let's tie it all together.

1 The oldest known fossils of living things look very similar to the cell walls of modern bacteria. How old are these earliest fossils of living cells?

2 You find a single-celled organism, and you want to determine whether it is a bacterium, archaean, or protist. Your uncle says it is a bacterium, for sure. What characteristics would you need to see to allow you to definitively determine if your uncle is correct?

3 How is the term "prokaryote" a deceptive, phylogenetic lie?

4 Why are "protists" not considered a natural group?

5 How does horizontal gene transfer and symbiogenesis complicate reconstructing and visualizing the tree of life?

6 How has Earth's atmosphere been altered by bacteria and archaea? (*Hint:* Think O_2.)

7 Using careful, sterile sampling techniques, your research team succeeds in drilling through a mile of ice and recovering a new species of bacterium from a subglacial lake in Antarctica. You determine that it acquires energy and carbon from inorganic methane seeping into the lake from the subsurface. What is the trophic category of the new species?

Exercise 27
Survey of Eukarya: Fungi and Lichens

Objectives

At the completion of this exercise, students will be able to:

1 Define key terms associated with fungi and lichens.
2 Describe the distinguishing characteristics and identify certain types of fungi and lichens.
3 Describe the ecological role of fungi.
4 Illustrate the basic parts of a typical mushroom.

ⓘ Background Information

Fungi are unique eukaryotes and, despite their similarity to plants, are more closely related to animals. There are over 100,000 described species of fungi. Representative taxa include mushrooms, puffballs, bread mold, morels, truffles, smuts, rusts, blight, mildew, and yeasts. Fungi are heterotrophic, filamentous, spore-producing, unicellular or multicellular eukaryotes.

Lichens, with few exceptions, are symbionts made up of a green algae or cyanobacteria and ascomycetes. It is believed that the algae or cyanobacterium partner in the relationship provides the food for the lichen through the process of photosynthesis. The ascomycete partner provides mineral and water retention and provides attachment and protection. Lichens can usually be found attached to trees, rocks, and man-made structures such as walls, sidewalks, and gravestones. There are three types of lichens: crustose, foliose, and fruticose.

Activity 1 Key Terms

Materials Required
❏ Biology textbook or access to internet resources

Define the following terms:

Eukarya

_____

_____

Symbiosis

_____

_____

Heterotrophic

_____

_____

Crustose lichen

_____

_____

Foliose lichen

_____

_____

Fruticose lichen

_____

_____

Zygomycete

_____

_____

Ascomycete

_____

_____

Basidiomycete

_____

_____

Materials Required
❏ Digital camera or sketchpad and pencil

1 With a digital camera, or a sketchpad and pencil, find, document, and identify the organisms in the list below. Most of these are common in and around your home, but you might be surprised what you can find at the grocery store as well.

a. Crustose lichen

b. Foliose lichen

c. Fruticose lichen

d. Zygomycete

e. Ascomycete #1

f. Ascomycete #2

g. Basidiomycete #1

h. Basidiomycete #2

i. Basidiomycete #3

2 Produce a diary or logbook with your photos or sketches. Label each photo or sketch with the corresponding letter from the list in step 1.

Activity 3 Synthesis

Materials Required
No materials required for this activity

Now let's tie it all together.

1 What important drugs come from mold?

2 What important food products come from mold/yeast?

3 What biomedical uses do lichens have?

4 How are lichens used to monitor air quality?

5 Describe the ecological role and functional characteristics of fungi.

6 How does lichen differ from fungi?

7 Is lichen a natural taxonomic/phylogenetic group? (*Hint:* Does it have a unique origin?) Explain.

Exercise 28
Survey of Eukarya: Plantae (Viridiplantae)

Objectives

At the completion of this exercise, students will be able to:

1 Define and apply key botanical terms.

2 Describe the phylogenetic position of plants on the tree of life.

3 Describe the four morphological differences between monocots and dicots.

4 Compare and contrast vascular and nonvascular plants.

5 Describe and sketch the basic leaf types.

6 Illustrate the basic parts of a flower and describe their role in reproduction.

ⓘ Background Information

Plants are multicellular eukaryotes in the kingdom Plantae. Green plants have cell walls that contain cellulose. Photosynthesis is the process by which plants obtain most of their energy. They have chloroplasts that contain a green pigment called chlorophyll.

There are thought to be more than 310,000 species of green plants. Plants provide most of the Earth's oxygen, and are the primary source of carbon and energy that run Earth's ecosystems and resulting ecosystem services. Humans rely on plants as their fundamental source of energy (food), medicine, fiber for clothing and construction materials, and as ornamental and landscape adornments.

Activity 1 Key Terms

Materials Required

❑ Biology textbook or access to internet resources

Define the following terms:

Fruit

_____

_____

Vegetable

_____

_____

Flower

_____

_____

Xylem

_____

_____

Phloem

Cambium

Dicot

Monocot

Eudicot

Vascular plant

Nonvascular plant

Gymnosperm

Angiosperm

Complete flower

Incomplete flower

Activity 2 Phylogenetic Position of Plantae (Viridiplantidae)

Materials Required

❑ Biology textbook or access to internet resources
❑ Digital camera or sketchpad and pencil

1 Using your textbook or internet resources as a guide, draw a phylogenetic tree that depicts the position of plants relative to the other major organismal lineages. In your tree include the following:

❑ Plants   ❑ Animals   ❑ Fungi   ❑ Bacteria   ❑ Archaea

2 Many scientific names are descriptive, and plants are no exception. Many taxonomic names, such as the genus and species, provide insight into their morphology. Other times they may be named after their geographic location, or the person who named them. In the space below, provide the common name, genus, and species of a plant you find particularly beautiful.

For example:

Common name: Sunflower

   Genus: *Helianthus*

      Meaning: Helios = sun

   Species: *annuus*

      Meaning: annual, yearly

Common name: _____

   Genus: _____

      Meaning: _____

      _____

   Species: _____

      Meaning: _____

      _____

Activity 3 Plant Structure

Materials Required
❏ Biology textbook or access to internet resources
❏ Colored pencils

1 Using your textbook or internet resources, illustrate the four differences between monocots and dicots.

2 Begin with monocot and dicot embryos. Draw a picture of a monocot embryo, and a picture of a dicot embryo. For each picture, be sure to label the cotyledons.

3 Look at monocot and dicot stem cross sections. Draw a picture of a monocot stem cross section, and a picture of a dicot stem cross section. For each picture be sure to illustrate the pattern of vascular tissue.

4 Note the veins in monocots and dicots. Draw a picture that illustrates the veins in a monocot leaf, and a picture of the veins in a dicot leaf.

5 Observe monocot and dicot flowers. Draw a picture that illustrates the differences in the numbers and arrangement of flower petals in monocots and dicots.

6 Using your textbook or internet resources, draw a picture that illustrates the basic parts of a flower. Include in your drawing: carpel, sepal, petal, stamen, stigma, style, anther, filament, ovary, and receptacle.

Activity 4 Exploring Plant Biodiversity

Materials Required
❏ Digital camera or sketchpad and pencil

1 With a digital camera, or a sketchpad and pencil, find, document, and identify (using common or scientific name, if known) representative organisms or examples from the list that follows in step 2.

2 Produce a field book or notebook with your photos or sketches and label each photo or sketch with the corresponding letter.

a. Nonvascular plant #1
b. Nonvascular plant #2
c. Seedless vascular plant #1
d. Seedless vascular plant #2
e. Gymnosperm #1
f. Gymnosperm #2

g. Dicot #1
h. Dicot #2
i. Monocot #1
j. Monocot #2
k. Modified root
l. Terminal bud
m. Simple leaf

n. Pinnate compound leaf
o. Palmate compound leaf
p. Leaf with parallel venation
q. Leaf with palmate venation

r. Specialized leaf
s. Complete flower
t. Incomplete flower
u. Berry
v. Pome
w. Aggregate fruit

Activity 5 Synthesis

Materials Required
No materials required for this activity

Now let's tie it all together.

1 What's the difference between a fruit and a vegetable?

_____

_____

_____

2 List the function of each of the following parts of a flower:

a. Ovary

_____

_____

b. Ovule

_____

_____

c. Anther

_____

_____

d. Stigma

_____

_____

3 Describe the events that occur during pollination.

_____

_____

_____

_____

_____

_____

_____

4 Provide 12 reasons why you should "thank a plant."

_____

_____

_____

_____

_____

_____

_____

_____

_____

_____

_____

Exercise 29
Survey of Eukarya: Metazoa

Objectives

At the completion of this exercise, students will be able to:

1 Define and apply key metazoan terms.

2 Describe the functional and ecological roles of metazoans.

3 Describe the key features of animals associated with the basic subgroups within the kingdom Animalia.

(i) Background Information

Animals are multicellular, eukaryotic organisms belonging to the kingdom Metazoa (also called Animalia) and occupy every niche found on all continents on earth. Kingdom Metazoa is divided into various subgroups, including vertebrates, molluscs, arthropods, annelids, sponges, and jellyfish.

All animals eventually achieve a fixed body plan as they develop. Some, however, undergo a metamorphosis, a process in which they develop into a different form, such as a caterpillar transforming into a butterfly, for example.

Animals are heterotrophic, meaning that they must ingest other organisms, parts of other organisms, or products of organisms for nourishment.

Activity 1 Key Terms

Materials Required
❏ Biology textbook or access to internet resources

Define the following terms:

Metazoan

_____

_____

Heterotrophs

_____

_____

Motile

_____

_____

Animalis (Latin)

_____

_____

Materials Required
☐ Digital camera or sketchpad and pencil

1 Using a digital camera, or a sketchpad and pencil, find, document, and identify the organisms from the list provided in step 2. You may find these organisms in the wild, zoo, wildlife park, museum, or aquarium.

2 Produce a diary or logbook with your photos or sketches and label each photo or sketch with the corresponding letter.

a. Annelid #1	h. Insect #1	o. Fish #2	v. Bird #3
b. Annelid #2	i. Insect #2	p. Amphibian #1	w. Mammal #1
c. Nematode	j. Insect #3	q. Amphibian #2	x. Mammal #2
d. Crustacean #1	k. Insect #4	r. Reptile #1	y. Mammal #3
e. Crustacean #2	l. Insect #5	s. Reptile #2	z. Mammal #4
f. Arachnid #1	m. Insect #6	t. Bird #1	
g. Arachnid #2	n. Fish #1	u. Bird #2	

Activity 3 Synthesis

Materials Required
No materials required for this activity

Now let's tie it all together.

1 Some animals seem to be more similar to plants than animals. This is especially true for the sessile ones—those that just sit in one place for the majority of their lives. For example, the giant clam *Tridacna* sits in one place all the time sunning itself and essentially farming photosynthetic algal symbionts from which it draws its nutrients. Sea anemones and corals just sit in the same place their whole lives catching organic material that falls from above. How is that any different from a pitcher plant? Answer the following questions to reach a conclusion:

a. Why do we still refer to a giant clam as a metazoan?

_____

_____

_____

b. How is a pitcher plant similar to a sea anemone?

_____

_____

_____

c. How is it different?

_____

_____

_____

_____

2 What's the difference between a metazoan and a protozoan?

_____

_____

_____

_____

Ecology

In This Section

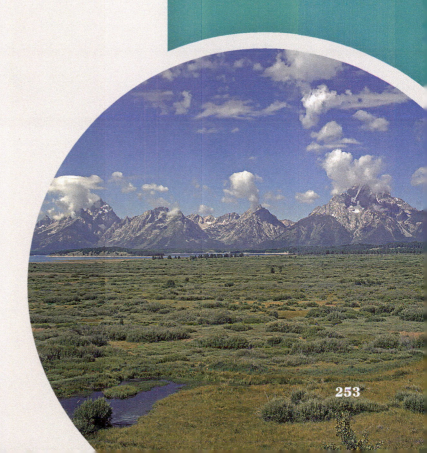

Exercise 30
Animal Behavior

Objectives

At the completion of this exercise, students will be able to:

1 Define and apply key terms associated with animal behavior.

2 Learn different methods for sampling animal behaviors.

3 Collect behavioral data and use them to generate ethograms.

4 Use ethograms to draw inferences about animal behaviors.

5 Develop field-based observational skills sufficient to effectively make inferences about animal behaviors.

6 Effectively communicate behavioral data graphically.

7 Generate a scientific question, a null, and an alternative hypothesis, and make corresponding predictions about animal behavior.

ⓘ Background Information

Various sampling methods are used by biologists studying animal behaviors, including the following:

Ad Libitum Sampling is a method in which the observer takes free-form notes on what is happening. This technique by definition is not systematic or quantitative and is most often used to get ideas and to begin to become familiar with an animal.

Focal Animal Sampling is the name given to systematic observation techniques in which the observer concentrates on just one individual.

All Occurrence Sampling is when all occurrences of a behavior are recorded as they occur.

Instantaneous ("On the Beep") Sampling is a method in which the behavior of a group or individual is recorded at set time intervals. The animal's behavior at the moment that the time interval expires is recorded.

Scan Sampling is sometimes used synonymously with instantaneous sampling. In scan sampling, a group or individual is scanned at set time intervals, and whatever the group or individual is doing at the moment of the scan is recorded.

Collecting behavioral data is fundamentally no different than any other scientific endeavor. Your data should be accurate, replicable, and directly relevant to the scientific question you are interested in addressing. As you are recording your observations, you should try to include every detail you can think of in your field notes, including the location, species involved, temperature, habitat, wind speed, weather conditions, etc. It is always preferable to have too many observations than too few (Fig. 30.1).

Species:	*Turdis migratorius*	Observer:	Bob
Date:	3-Dec-13	Time:	14:00–15:00
Conditions:	Light snow falling, 28°F, light winds	Comments:	Adult male
Locations:	City Park		

Time	Category of Behavior				
	Foraging	**Singing**	**Stationary**	**Flying**	**Walking**
14:02	X				
14:04	X				
14:06		X			
14:08		X			
14:10		X			
14:12		X			
14:14			X		
14:16			X		
14:18				X	
14:20			X		
14:22				X	
14:24		X			
14:26					
14:28					X
14:30	X				X
14:32	X				

FIGURE **30.1** ● Sample data sheet.

Once you have completed your sampling, you will be able to create an ethogram. An ethogram is a summation of the behaviors that are carefully described and organized into categories. Ethograms place the repertoire of behaviors into an organized framework that facilitates descriptions of how each behavior is involved in survival, reproduction, feeding, mating, etc. They are usually depicted graphically, as shown in Figure 30.2.

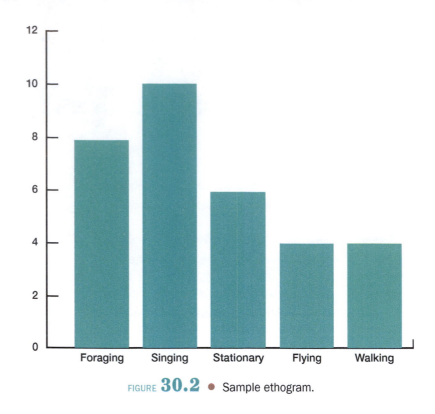

FIGURE **30.2** ● Sample ethogram.

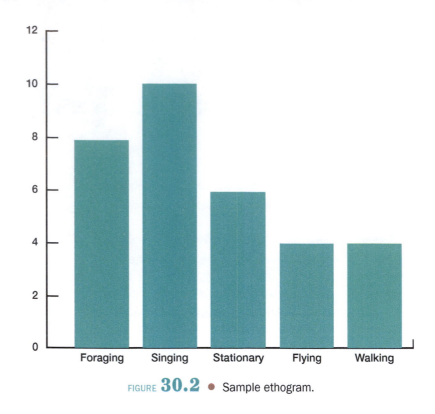 **Activity 1** Key Terms

Materials Required
❏ Biology textbook or access to internet resources

Define the following terms:

Ethogram

_____

_____

Experimental unit

_____

_____

Behavioral ecology

_____

_____

Evolutionary Stable Strategy (ESS)

_____

_____

Cooperation

Kin recognition

Kin selection

Competition

Reciprocal altruism

Sexual selection

Parental care

Mating systems

Inclusive fitness

Sexual conflict

Activity 2 Scientific Contributions

Materials Required
❑ Biology textbook or access to internet resources

1 Describe the contributions that Niko Tinbergen and Konrad Lorenz made to the understanding of animal behavior patterns.

_____

_____

_____

_____

_____

Activity 3 Observations of Animals

Materials Required
❑ Pencil or pen for taking notes and recording observations
❑ Access to observable animals

1 Observe the behavior of any (nonhuman) animal species. Use any one of the sampling methods described at the beginning of this exercise.

2 Decide *a priori* (without any prior experience or knowledge) which four behaviors you will measure. Record behavioral responses, taking care not to anthropomorphize (e.g., "robins sing because they are happy"), and fill out Table 30.1. The following are some suggestions for observing animal behavior:

☐ Go to a park to observe songbirds or waterfowl.

☐ Look out your window at a bird feeder.

☐ Observe cows, sheep, goats, etc., in a pasture.

☐ Watch houseflies in your kitchen.

☐ Monitor ants in your backyard.

TABLE **30.1** Behavioral Data

Species:	Observer:
Date:	Time:
Conditions:	Comments:
Location:	

Time	Behavior 1	Behavior 2	Behavior 3	Behavior 4	

3 Using Figure 30.3, construct an ethogram using your observational data.

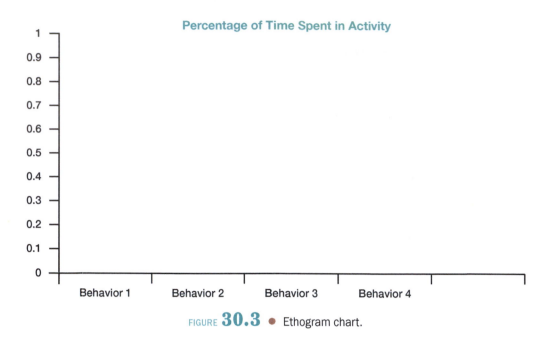

FIGURE **30.3** ● Ethogram chart.

 **Activity 4** Synthesis

Materials Required
No materials required for this activity

Now let's tie it all together.

1 What new question about animal behavior arises from your observations?

_____

_____

_____

2 Describe your null and alternative hypotheses.

_____

_____

_____

3 What do your hypotheses predict?

_____

_____

_____

_____

4 How could you use a phylogeny to make predictions about the origin and maintenance of the behaviors you observed?

5 Humans are animals too.

   a. Why do we distinguish between human behavior and animal behavior?

   b. How might studying human behavior be similar to studying animal behavior?

   c. How might studying human behavior be different from studying animal behavior?

   d. What ethical questions might arise when studying nonhuman animals?

   e. What ethical questions might arise when studying human behavior?

Exercise 31
Animal Physiology

Objectives

At the completion of this exercise, students will be able to:

1 Define and apply key physiological terms.
2 Predict the effect of the environment (temperature) on ectotherm physiology.
3 Describe the effect of temperature on physiological respiration rates.
4 Design an experiment to explore the effect of temperature on behavior.
5 Distinguish acclimation from acclimatization.

ⓘ Background Information

Goldfish, *Carassius auratus*, are one of the earliest fish species to be domesticated. They are thought to have originated in China over a thousand years ago. They are perhaps the most common aquarium fish available today. Goldfish thrive in cold water and can survive through winter in ponds that freeze over. They range in size from small aquarium fish to individuals up to 18 inches long and over 4 pounds in weight.

Physiological respiration is the process by which organisms exchange gases with their environment. Aerobic respiration, or aerobic metabolism, occurs in the animal kingdom when oxygen is taken into the body and distributed to all its cells where it is used to help process food into energy. Respiration in goldfish may be observed by watching their mouth. Goldfish take water in through their mouth and send it out over their gills and through the opercula. To determine the respiration rate, one only needs to count the number of "breaths" a goldfish takes in a given time.

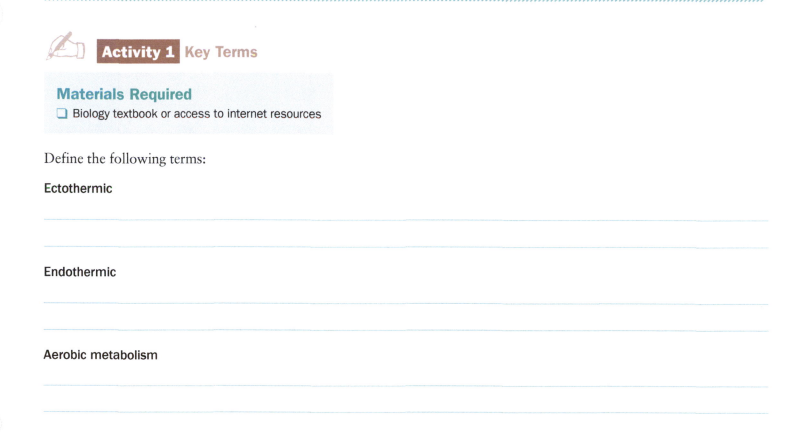

Activity 1 Key Terms

Materials Required
❑ Biology textbook or access to internet resources

Define the following terms:

Ectothermic

_____

_____

Endothermic

_____

_____

Aerobic metabolism

_____

_____

Acclimation

_____

_____

Acclimatization

_____

_____

 Activity 2 Effect of Temperature on Physiological Respiration

Materials Required

- ❏ Aged water at room temperature
- ❏ Ice (1 cup)
- ❏ Clear jar capable of holding at least 6 or 7 cups of water
- ❏ Goldfish
- ❏ Wristwatch or other timing device

1 The goldfish is an ectothermic animal, meaning that it depends on its environment to regulate body temperature. Different temperatures will have different effects on the goldfish. What will happen to the respiration of a goldfish as the water it's in is cooled down? Explain.

_____

_____

_____

2 Clean out a clear bottle or jar well and make sure there is no soap or other residues left.

3 Fill a second clean bottle with at least a half-gallon of water and let it stand for 24 hours at room temperature. Aging water allows chlorine and other harmful additives that may be in the water to break down or neutralize.

4 Once you have aged your water, acquire at least one goldfish from a local pet shop.

5 Pour 4 cups of aged water into your cleaned clear jar.

6 Allow the bagged goldfish to sit in aged water while the water temperature in the bag matches the water temperature in the jar (approximately 30 minutes).

7 After 30 minutes, release the goldfish and allow it to swim freely in the jar.

8 Once the goldfish has acclimated (10 minutes or so), count the number of "breaths" it takes over a one-minute period. Do this several times, and establish an average count.

9 Even though the goldfish thrives in cold water, sudden temperature changes can shock the fish. To avoid shocking the fish, place one cup of ice in the jar and allow it to melt on its own. This will cool the water but not all at once.

10 Once the ice has melted, observe the goldfish and count the number of "breaths" it takes over a one-minute period. Do this several times and establish an average count.

11 Use Figure 31.1 to graph your data. On the y-axis plot the temperature, and on the x-axis plot the respiration rate.

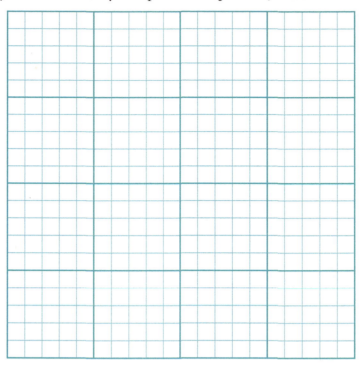

FIGURE **31.1** ● Graph for the effect of temperature on physiological respiration.

 **Activity 3** Synthesis

Materials Required
No materials required for this activity

Now let's tie it all together.

1 At which temperature did the goldfish have a higher respiration? Explain.

_____

_____

_____

2 Why does respiration change with temperature?

_____

_____

_____

3 Did your goldfish acclimate or acclimatize? Explain.

_____

_____

_____

4 Design an experiment to explore the effects of temperature on fish behavior.

Null Hypothesis: _____

_____

Alternative Hypothesis: _____

_____

Control group: _____

_____

Treatment group(s): _____

_____

How many experimental units (fish) do you need?

_____

_____

What is the response variable?

_____

_____

How will you measure the response variable?

_____

_____

_____

_____

What is the reasoning behind your predicted response?

_____

_____

_____

_____

Exercise 32
Population Ecology

Objectives

At the completion of this exercise, students will be able to:

1 Define and apply key terms related to population ecology.

2 Generate graphical reconstructions of population growth equations.

3 Calculate and plot population growth rates.

4 Compare and contrast the attributes and forces that shape arithmetic (linear), exponential, and logistic growth curves.

5 Apply calculations of growth rates to human populations.

6 Identify density-dependent and density-independent factors that influence population dynamics.

ⓘ Background Information

How does the number of individuals in populations change over space and time? How does birth rate, death rate, immigration, and emigration affect the overall growth of a population? Which biotic and abiotic drivers influence changes in the abundance and distribution of populations? What is the role of life history traits on population dynamics? At the root of each of these questions is the conclusion that Thomas Malthus arrived at in his 1798 paper, *An Essay on the Principle of Population,* which states that a population will grow exponentially so long as the environment remains constant. Of course, in the real world, resources and energy are limited and environments are changing; therefore, constraints are placed on the growth rates of populations of individuals competing for resources and space. Because human activities are the largest threat to biodiversity and the integrity of Earth's ecosystems, it is critical that we understand and address the drivers of human population dynamics. Similarly, the tools of population ecology are also indispensible for designing conservation management strategies for endangered populations, species, communities, and even ecosystems.

✍ Activity 1 Key Terms

Materials Required
☐ Biology textbook or access to internet resources

Define the following terms:

Population

_____

_____

Population ecology

_____

_____

Population density

_____

_____

Metapopulation

Geographic distribution (range)

Immigration

Emigration

Demography

Generation (time)

Life table

Survivorship

Survivorship curve

Type I curve

Type II curve

Type III curve

_____

_____

Fecundity

_____

_____

Age-specific fecundity

_____

_____

Fitness trade-offs

_____

_____

Per capita rate of increase (r)

_____

_____

Intrinsic rate of increase (r_{max})

_____

_____

Biotic potential

_____

_____

Arithmetic (linear) growth (+ equation)

_____

_____

Exponential population growth (equation)

_____

_____

Exponential population growth (density independent growth)

_____

_____

Exponential population growth (density dependent growth)

Carrying capacity (*K*)

Logistic growth equation

Lotka-Volterra equation

Population cycles

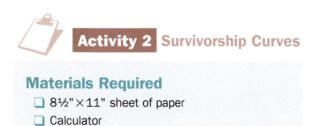

Activity 2 Survivorship Curves

Materials Required
- ☐ 8½" × 11" sheet of paper
- ☐ Calculator
- ☐ Colored pens, pencils, or markers

1 In Activity 1 you defined three types of survivorship curves, where mortality is low, steady, or high. In the plot in Figure 32.1, graph each of the three types of curves. Below each curve provide an example of an organism that exhibits each type of survivorship.

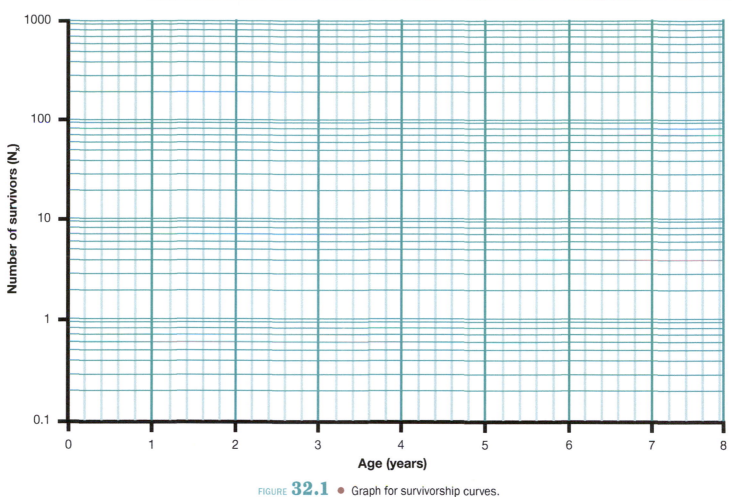

Number of survivors (N_x)

Age (years)

FIGURE **32.1** • Graph for survivorship curves.

Activity 3 Demonstrating Exponential Growth

Materials Required
- ❑ 8½" × 11" sheet of paper
- ❑ Calculator
- ❑ Colored pens, pencils, or markers

This activity is a terrific bar bet, but an even better illustration of the counterintuitive (for many) concept of exponential growth. The next time you're at a party, and the conversation turns a bit dull, liven it up by challenging all comers to fold a piece of paper eight times. They'll look foolish while you collect on your bet! Try this exercise now yourself, and you'll understand why it works.

1 How many times do you think you can fold a single 8½" × 11" sheet of paper in half?

_____

2 Try to see how many times you can fold the paper in half (consecutively—you can't just fold it and unfold it eight times!). How many times were you able to fold it in half?

_____

3 Why did you think it wouldn't be that hard to fold a piece of paper in half eight times?

_____

_____

_____

4 Fill out Table 32.1 that shows how many layers of paper you get with each fold.

TABLE **32.1** Number of Folds vs. Layers of Paper (Thickness)

Number of Folds	Layers of Paper (Thickness)
0	1
1	2
2	4
3	
4	
5	
6	
7	
8	
9	
10	
11	
12	
13	

5 Graph your results using the graph in Figure 32.2. Graph the number of folds on the x-axis and the number of layers of paper (thickness) on the y-axis. Before you start graphing, look at the data in your table. You're going to need to choose your scales carefully, because by the time you get to 10 folds, your paper is already more than 1,000 layers thick!

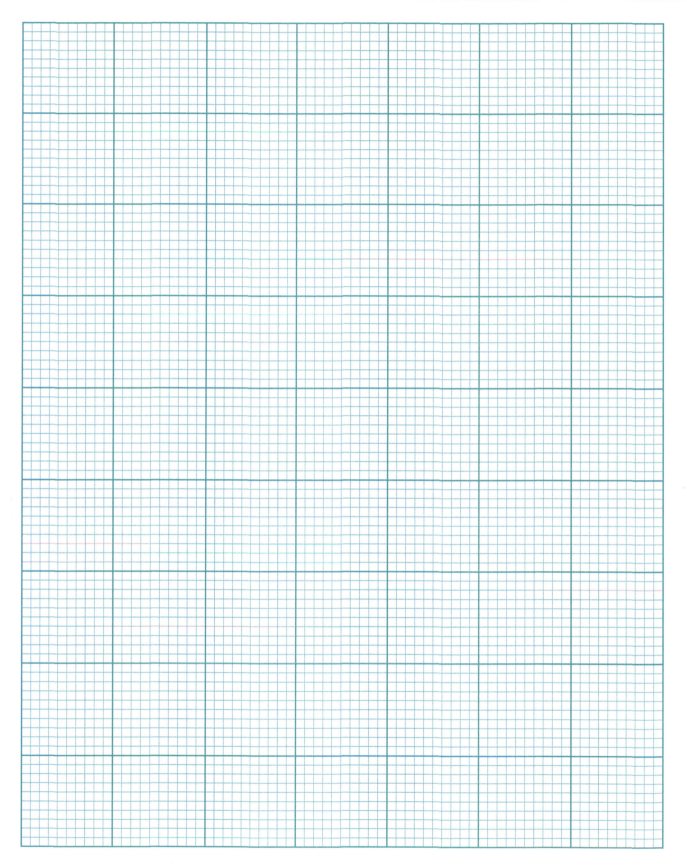

FIGURE **32.2** ● Graph for number of folds vs. layers of paper (thickness).

6 What pattern do you see developing in your graph as you increase the number of folds?

_____

_____

_____

7 Why does it start slow, and then increase so rapidly? (***Hint:*** The base of what is multiplied?)

_____

_____

_____

_____

Activity 4 Calculating Exponential Growth

Materials Required
☐ 8½" × 11" sheet of paper
☐ Calculator
☐ Colored pens/pencils/markers

1 A female fruit fly, _Drosophila melanogaster_, can lay up to 100 eggs per day and as many as 2,000 eggs over her lifetime. For ease in calculation, let's assume that a female fruit fly lays 100 eggs every two weeks, and that this works out to be 25 generations per year. Assuming that all of the offspring survive and reproduce at the same rate as their parents (100 eggs every two weeks), how many flies will you have at the end of one year (25 generations)? To calculate this, use Table 32.2 below. We start the first generation with two flies (one male and one pregnant female). By the end of the second generation, the two flies will have had 25 generations of 100 eggs each, yielding 5,000 flies by the start of the third generation.

TABLE **32.2** Fruit Fly Generations

Number of flies	2	100	5,000			
Number of generations	1	2	3	4	5	25

   Fruit flies are pretty small at only 3mm long, 2mm wide, and 2mm high. How much space would this many flies occupy? That's 12 cubic mm per fly, so after 25 generations you end up with 1.428×10^{39} cubic meters of fruit flies, which would be a giant ball of fruit flies nearly the volume of Jupiter (1.4313×10^{15} km³)! Thankfully, nature imposes constraints on survival and successful reproduction in the real world that would prevent this from ever happening.

2 Let's look at Earth's human population in Table 32.3. Using the graph in Figure 32.3, plot these values with year on the x-axis and population on the y-axis.

TABLE **32.3** Human Population of Earth

World Population Reached	Year	Time to Add 1 Billion
1 billion	1804	
2 billion	1927	123 years
3 billion	1960	33 years
4 billion	1974	14 years
5 billion	1987	13 years
6 billion	1999	12 years

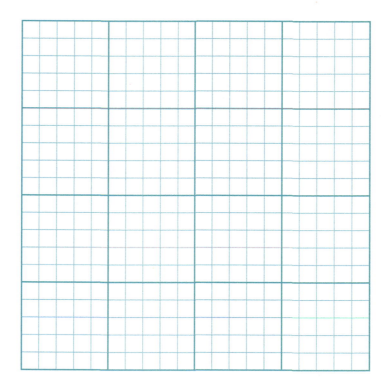

FIGURE **32.3** ● Graph of population of the Earth.

3 How is the curve different from the unregulated population growth for fruit flies?

_____

_____

_____

_____

4 What trend do you notice regarding the length of time it takes to add a billion people to the planet?

 Activity 5 Everyday Exponential Growth: A More Realistic Example

Materials Required
- ❏ 8½" × 11" sheet of paper
- ❏ Calculator
- ❏ Colored pens/pencils/markers

Population growth rate is the rate at which a given number of individuals in a population grows (increases) over a given period of time in proportion to the initial population. In the paper-folding example, the growth rate was 100 percent for each event, but in the real world, biological populations rarely grow at this rate. For example, human population growth rate is currently 1.1 percent per year. Deer populations in parts of North America can grow at a rate of 10 percent per year, and the growth rate of cockroaches can be as high as 10 percent per month.

1 To calculate the population growth rate of any species, take the difference between the birth rate and death rate, and add it to the difference between the immigration rate and the emigration rate, using the following equation:

$$N = B - D + I - E$$

2 Assume a certain town in your area has 1,000 people. Last year, 16 new babies were born and 12 people died, but nobody moved into the population and nobody moved out. Sounds like the population is growing, but at what rate? (***Note:*** $N = B - D/1000 \times 100$.)

3 In the exercise below, assume you moved into an apartment that has 25 cockroaches living in your kitchen. Use the cockroach growth rate to calculate the expected population size after two years (24 months). For example, to calculate how many roaches you have in your kitchen by the second month, you multiply the initial number of cockroaches by 10% (0.10), and add that value to the start population:

$$25 \times 0.1 = 2.5;$$ so after two months you have 30.25 cockroaches, and so on.

4 Annotate where in Table 32.4 the population doubles.

5 What pattern of doubling time emerges from your data?

TABLE **32.4** Number of Cockroaches

Time (Months)	Number of Cockroaches
0	25
1	27.5
2	30.25
3	
4	
5	
6	
7	
8	
9	
10	
11	
12	
13	
14	

Activity 6 Comparison of Arithmetic Growth versus Exponential Growth

Materials Required
No materials required for this activity

Can food production keep pace with reproduction? Food production can benefit from technological advances and efficiencies, but can optimistic growth in crop yields keep pace with human population growth? To answer this question, you need to understand the difference between exponential growth and arithmetic growth.

1 Complete the missing values for the years 2021–2036 in Table 32.5. To calculate the missing values, note that increases in yield are arithmetic, whereas gains in population are exponential. Use the arithmetic and exponential growth equations you defined in Activity 1.

TABLE **32.5** Arithmetic Growth

Year	Yield	Population Size
2016	100	100
2017	105	105
2018	110	110
2019	115	116
2020	120	122
2021	125	128
. . .		
2026		
. . .		
2036		

2 Graph the curve for food production and population growth on the graph in Figure 32.4.

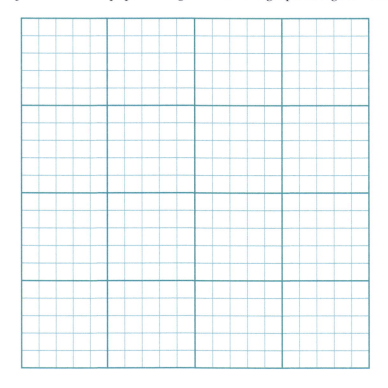

FIGURE **32.4** ● Graph for food production and population growth.

3 What assumptions are made about the human population rate increase?

_____

_____

_____

4 What assumptions are made about the rate of food production?

_____

_____

_____

Activity 7 Carrying Capacity and the Logistic Growth Function

Materials Required
❑ Red and green pencils or markers

As with the cockroaches, human populations cannot continue to grow exponentially; at some point, we will reach the carrying capacity of the planet. Thus, the growth curve of all populations (humans included) will eventually approximate a logistic curve. That is, the population will increase exponentially until it reaches a point at which individuals in the population are affected by environmental constraints on their growth. At that point, the rate of growth declines until the population reaches carrying capacity, and the growth rate of the population reaches zero.

1 Using the area of the graph in Figure 32.5, draw a picture of a logistic growth curve.

2 Use a red pencil or marker to color the part of the curve that depicts the initial phase (early, rapid, exponential growth). Use a green pencil or marker to color the later part of the curve showing the decline in growth rate (starts at the sigmoidal midpoint, which is the midpoint of the S-shaped curve of your graph).

3 Plot population size on the y-axis and time on the x-axis. Be sure to label your axes.

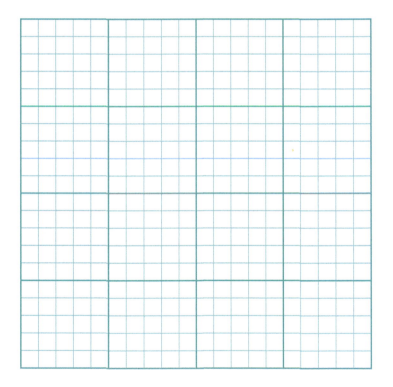

FIGURE **32.5** ● Graph of a logistic growth curve.

4 List some density-independent factors that could be responsible for the shape of this curve. Consider some abiotic factors that could potentially lead to decreased birth rates or increased death rates?

_____

_____

_____

_____

5 List some density-dependent factors that could be driving the shape of the logistics growth curve. (**Hint:** Consider what biotic factors could be driving changes in birth rates or mortality?)

_____

_____

_____

_____

 Activity 8 Population Dynamics: The Lotka-Volterra Equation

Materials Required
No materials required for this activity

When two species affect each other's population dynamics, their growth curves become interdependent. For example, in mutualistic relationships, the growth curves of both species will be very similar. This is also true if two species compete for the same resource. Similarly, the curve of a commensal organism (an organism that benefits from another organism without affecting it) will closely track that of its host (but not vice versa). Predator-prey (and host-parasite) growth curves will cycle; the crash in the prey/host population is usually followed by a crash in the predator population. Thus, they are synchronous but have a lag effect by the predator, parasite, or pathogen.

1 In the graph in Figure 32.6, use your textbook or internet resources to reconstruct a graphical representation of the population dynamics of the Canadian snowshoe hare and its predator, the lynx. On the y-axis depict their population density, but keep in mind that these will likely be at different scales! On the x-axis depict time. Draw the rabbit curves in red and the lynx curves in blue. Include at least a 40-year span on the x-axis. Be sure to label the axes and main features of your graph.

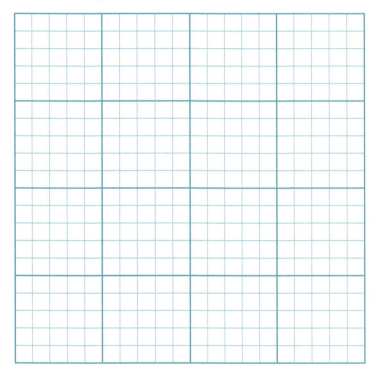

FIGURE **32.6** ● Graph of the population dynamics of the Canadian snowshoe hare and the lynx.

2 What is the primary density-dependent factor driving the pattern of lynx populations?

_____

_____

3 What are some density-independent explanations for changes in the hare abundance?

_____

_____

_____

4 How might density-independent changes to the environment, such as climate change, affect the population density of the lynx in the future?

_____

_____

_____

_____

_____

_____

Activity 9 Synthesis

Materials Required
No materials required for this activity

Now let's tie it all together.

1 What are the four factors that define the growth rate of a population?

_____

_____

_____

_____

2 *Plasmodium* is a vector for mosquitoes. It is a protozoan that is the causal agent of malaria, which is responsible for killing as many as 855,000 people in 2013, the majority of them being children. Populations of mosquitoes are strongly influenced by the availability of standing water for egg-laying and larval development. How might mosquito density, and thereby human mortality, be affected by patterns of rainfall?

3 What are some factors that cause a population to reach carrying capacity?

4 What is the likelihood that Earth's population problems will be solved by technological advancements? Explain.

5 What kind of environmental policy message emerges from what you know about the difference between arithmetic (linear) growth and exponential growth?

Exercise 33
Ecological Succession

Objectives

At the completion of this exercise, students will be able to:

1 Define and apply key terms related to ecological succession.

2 Explain the progressive stages that forests and ponds go through by describing the composition of different species and densities prior to reaching a climax community.

3 Predict the changes that ecosystems will undergo based on their present condition and climate trends.

4 Identify successional processes in their local environment

ⓘ Background Information

Ecological succession is a predictable pattern of changes that takes place in all ecosystems in response to disturbance or initial colonization of new habitat. The patterns of succession appear to be common, even among very different environments. However, scientists have yet to determine if the processes that produce these patterns are driven by the same, or different mechanisms. In general, the first phases of succession start when new pioneering organisms colonize an unoccupied habitat following disturbance, such as lava flows and mudslides (primary succession), or fire, flooding, logging, mining, and converting prairies, forests, and deserts for agricultural purposes (secondary succession). These stages can often take many hundreds of years to complete and are difficult to observe directly. It is important to know and understand the different stages so that we can better understand and predict how ecosystems will respond to future environmental and ecological changes.

✍ Activity 1 Key Terms

Materials Required
☐ Biology textbook or access to internet resources

Define the following terms:

Succession

_____

_____

_____

Seral stages

_____

_____

_____

Climax community

_____

_____

_____

Disturbance

Stages of succession

Initial

Intermediate

Subclimax

Climax

Primary succession

Secondary succession

Ecotone

Pioneer species

_____

_____

Non-pioneer species

_____

_____

Shade-tolerant

_____

_____

Shade-intolerant

_____

_____

Facilitation

_____

_____

Competition

_____

_____

Materials Required

❑ Biology textbook or access to internet resources

As the last ice age ended, the water levels of larger lakes were significantly higher. As the ice receded and the water levels of large lakes fell, more land was exposed, and numerous smaller lakes and ponds where formed. Over the decades and centuries, these lakes and ponds turned to marshes and some eventually become forests (Fig. 33.1).

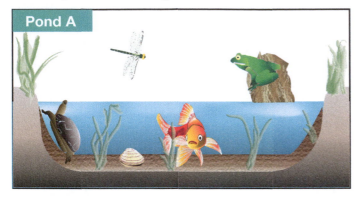

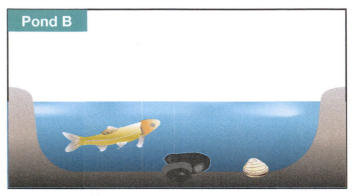

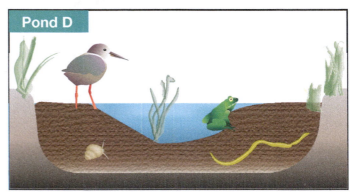

FIGURE **33.1** ● Stages of pond formation.

1 Using the internet and other available resources, list the animal and plant species most likely to inhabit each of the stages of pond formation shown in Figure 33.1. Using key terms from this lesson, provide information supporting your answers.

a. Pond A

b. Pond B

c. Pond C

d. Pond D

2 Using the previous four representative stages of ecological succession in a pond environment, list the stages in the appropriate order and explain your answer.

3 Another series of stages of succession take place after a disturbance such as a forest fire. Using the resources available to you and Figure 33.2, use key terms from this lesson to explain what is taking place.

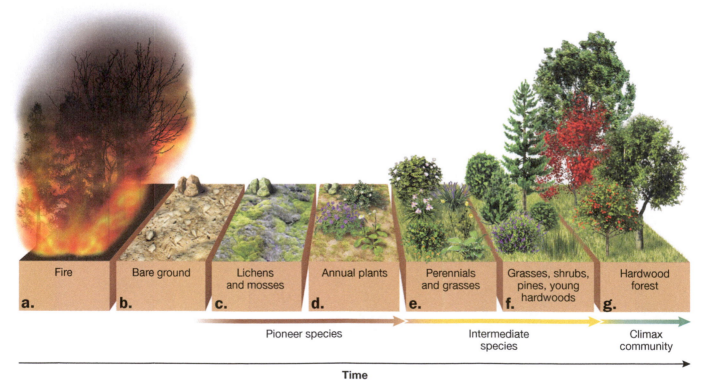

FIGURE **33.2** ● Climax community timeline.

4 Figure 33.2 illustrates floral (plant) succession. For each step in floral succession describe the animals that you might expect to find.

a. _____

_____

b. _____

_____

c. _____

_____

d. _____

_____

e. _____

_____

f. _____

_____

g. _____

_____

Activity 3 Synthesis

Materials Required
No materials required for this activity

Now let's tie it all together.

1 Make a list of all the ecological disturbances that you can think of. Include those that occur across very small to very large spatial scales. Which lead to primary succession? Which lead to secondary succession?

2 Choose one of your disturbances that leads to secondary succession and answer the following questions:

   a. What is the effect of the disturbance on the species diversity?

   b. To which type of primary disturbance is it most similar?

3 What role does random chance play in succession?

4 What role does dispersal play in succession?

5 What role does shade tolerance play in succession?

Exercise 34
Biogeochemical Cycles

Objectives

At the completion of this exercise, students will be able to:

1 Define and apply key terms related to biogeochemical cycles.
2 Discuss the components of biogeochemical cycles and how nutrients move through the biosphere.
3 Discuss how all biogeochemical cycles are connected, either directly or indirectly.

Part 1: The Water Cycle

(i) Background Information

More than 97 percent of the Earth's water is salt water, and most of the remaining freshwater (<2.5 percent) is locked up in ice caps and glaciers. Only about 0.5 percent of Earth's freshwater is in available from lake, river, and groundwater sources. Most water vapor moves into the atmosphere through evaporation. As it cools, the water vapor condenses, forming clouds, and then falls back to the earth as precipitation such as rain and snow. Most precipitation occurs over the ocean, but precipitation that does fall on land seeps into the ground and is stored in aquifers or forms lakes and rivers that eventually lead to the ocean. Once diminished, some aquifers can take thousands of years to recharge. In cold regions, some of the snow that falls forms glaciers. A drop of precipitation that lands on a glacier can spend over a million years in a glacier, or thousands of years in the ocean, before moving on to another part of the water cycle. Conversely, a water molecule only lasts about 8 days in the atmosphere before returning to the Earth's surface.

Activity 1 Key Terms

Materials Required
☐ Biology textbook or access to internet resources

Define the following terms as each relates to the water cycle.

Evaporation

_____

_____

Condensation

_____

_____

Precipitation

_____

_____

Surface runoff

_____

_____

Subsurface runoff

_____

_____

Percolation

_____

_____

Aquifer

_____

_____

Snow and ice

_____

_____

Ground water

_____

_____

Soil moisture

_____

_____

Activity 2 Sketch of the Water Cycle

Materials Required
- ❏ Colored pencils
- ❏ Markers or crayons
- ❏ Blank sheets of unlined paper
- ❏ Access to the internet or textbook with diagrams of water and nutrient cycles

1 Using each of the key terms you defined in Activity 1, sketch and label the water cycle in the space provided. Include arrows indicating the direction of movement.

Activity 3 Synthesis

Materials Required
No materials required for this activity

Now let's tie it all together.

1 How do asphalt and impermeable concrete affect the global water cycle?

_____

_____

_____

2 How does the conversion of forests and prairies to agricultural fields affect the global water cycle?

_____

_____

_____

3 How does removal of water from aquifers affect the global water cycle?

_____

_____

_____

Part 2: The Nitrogen Cycle

(i) Background Information

The main reservoir of nitrogen is atmospheric N_2, a biologically inert diatomic molecule that makes up 78 percent of the air we breathe. It is only biologically useful when it is reduced, or "fixed," into ammonium (NH_4+) or nitrate ions (NO_3-). Fixation occurs naturally through lightning-driven reactions in the atmosphere and enzymatic activity of bacteria. However, through the production of industrial-based fertilizers, cultivation of crops that harbor nitrogen-fixing bacteria, and the emission of nitric oxide that results from the burning of fossil fuels, humans fix almost as much nitrogen as all other natural sources combined.

Activity 1 Key Terms

Materials Required
- ❑ Biology textbook or access to internet resources

Define the following terms as each relates to the nitrogen cycle:

Soil

_____

_____

Ocean

_____

_____

Atmosphere

_____

_____

Groundwater

_____

_____

Surface water

_____

_____

Rain

_____

_____

Plants

_____

_____

Fertilizer

_____

_____

Land animals

_____

_____

Land plants

_____

_____

Aquatic animals

_____

_____

Aquatic plants

_____

_____

Animal waste

_____

_____

Plant litter

_____

_____

Dead animals

_____

_____

Industrial fixation

_____

_____

Lightning

Nitrogen-fixing bacteria in soil

Nitrogen-fixing bacteria in sediments

Nitrogen-fixing cyanobacteria

 **Activity 2** Sketch of the Nitrogen Cycle

Materials Required
- ❏ Colored pencils
- ❏ Markers or crayons
- ❏ Blank sheets of unlined paper
- ❏ Access to the internet or textbook with diagrams of nutrient cycles

1 Using each of the key terms you defined in Activity 1, sketch the nitrogen cycle in the space provided. Include arrows indicating the direction of movement.

___

Activity 3 Synthesis

Materials Required
No materials required for this activity

Now let's tie it all together.

1 What are the main reservoirs of nitrogen?

_____

_____

2 What are the effects of cultivating crops that harbor nitrogen-fixing bacteria on the nitrogen cycle?

_____

_____

_____

_____

3 What are the effects of burning fossil fuels, which releases nitric oxide, on the global nitrogen cycle?

_____

_____

_____

_____

4 How does nitrogen runoff affect aquatic ecosystems?

5 What are some of the effects of increased nitrogen on the biosphere?

Part 3: The Carbon Cycle

ⓘ Background Information

Although the ocean is the largest carbon reservoir, the atmosphere is also a very important reservoir because carbon can move in and out of it so quickly. CO_2 is a greenhouse gas that traps heat on Earth. Photosynthesis pulls carbon (in the form of carbon dioxide, CO_2) out of the atmosphere and incorporates it into plant tissue. Conversely, cellular respiration releases the carbon that has been incorporated into the tissues of living organisms back into the atmosphere. Burning fossil fuels rapidly releases carbon that has been locked up in an inactive geological reservoir for millions of years into the atmosphere.

Activity 1 Key Terms

Materials Required
❏ Biology textbook or access to internet resources

Define the following terms as each relates to the carbon cycle:

Sunlight

Photosynthesis

Net primary productivity (NPP)

Automotive and industrial fossil fuel emissions

_____

_____

Plant respiration

_____

_____

Soil and root respiration

_____

_____

Animal respiration

_____

_____

Organic carbon

_____

_____

Marine organisms

_____

_____

Terrestrial organisms

_____

_____

Decay organisms (decomposers)

_____

_____

Dead organisms and waste products

_____

_____

Fossils and fossil fuels reservoir

_____

_____

Ocean reservoir

_____

_____

Atmospheric reservoir

_____

_____

Terrestrial reservoir

_____

_____

_____

Activity 2 Sketch of the Carbon Cycle

Materials Required
- ❑ Colored pencils
- ❑ Markers or crayons
- ❑ Blank sheets of unlined paper
- ❑ Access to the internet or textbook with diagrams of nutrient cycles

1 Draw a sketch depicting the carbon cycle in the space provided. Include arrows indicating direction of movement. Label each of the key terms you defined in Activity 1.

Activity 3 Synthesis

Materials Required
No materials required for this activity

Now let's tie it all together.

1 What is the largest carbon reservoir?

_____

2 How does carbon leave the atmosphere?

_____

3 How could increased atmospheric CO_2 affect global plant growth on land, and phytoplankton in the ocean (Net Primary Productivity; NPP)?

_____
_____
_____
_____

4 What happens to pH when CO_2 is absorbed by the ocean?

_____
_____

5 List some examples of how individual organisms are affected by changes in biogeochemical cycles.

_____
_____
_____
_____
_____

6 Many economists consider GDP (Gross Domestic Product) to be a key indicator of economic health and performance. How does NPP influence GDP? Why might NPP be a better index of human well-being than GDP?

_____
_____
_____
_____
_____

Part 4: Biogeochemical Cycles

ⓘ Background Information

All of the atoms that make up living organisms are part of biogeochemical cycles. Some of the most important of these atoms are nutrients—elements that are critical for metabolism, growth, and reproduction. These atoms can also be abiotic and present in water, air, and rocks. The places they are stored are called reservoirs. Reservoirs can store nutrients for different periods of time—some very quickly (minutes), others for thousands or millions of years. The same atoms are recycled over and over, and the paths they take as they move from abiotic systems through biotic systems is referred to as its biogeochemical cycle. The water, carbon, and nitrogen cycles from Parts 1 through 3 are biogeochemical cycles that are critical for sustaining life on earth. Humans have a tremendous influence over biogeochemical cycles resulting in significant changes to these cycles on a global scale.

✎ **Activity 1** Key Terms

Materials Required
☐ Biology textbook or access to internet resources

Define the following terms:

Biosphere

_____

_____

Lithosphere

_____

_____

Atmosphere

_____

_____

Hydrosphere

_____

_____

Activity 2 Seeing the Big Picture

Materials Required
- ❏ Colored pencils
- ❏ Markers or crayons
- ❏ Blank sheets of unlined paper
- ❏ Access to the internet or textbook with diagrams of nutrient cycles

1 Choose a study organism (any living organism on earth).

2 Describe how populations of this organism will respond to human-driven changes in the water, nitrogen, and carbon cycles.

3 Sketch a diagram to illustrate the magnitude of these changes over time. Include changes in geographic distribution, abundance, trophic interactions in food webs, and consequent impacts that these changes have on other species in its ecosystem, and the biosphere at large.

Activity 3 Synthesis

Materials Required
No materials required for this activity

Now let's tie it all together.

1 How are the water, carbon, and nitrogen cycles connected?

_____

_____

_____

_____

2 How are humans altering global water, nitrogen, and carbon cycles?

_____

_____

_____

_____

3 What are some of the effects of human-altered biogeochemical cycles on the biosphere?

_____

_____

_____

_____

4 How could global warming affect the water cycle?

_____

_____

_____

_____

Part 5: Ocean Acidification

ⓘ Background Information

About 25 percent of atmospheric CO_2 is absorbed by the ocean. When CO_2 dissolves it reacts with H_2O to form carbonic acid (HCO_3-) (Fig. 34.1).

$$CO_2 + H_2O + CO_3^{2-} \rightarrow 2HCO_3^-$$

FIGURE **34.1** ● Reaction of carbon dioxide with water.

The consumption of carbonate ions impedes the ability of organisms like corals, molluscs, and echinoderms to build calcium carbonate skeletons. This activity demonstrates how readily CO_2 is dissolved in water across a concentration gradient to form carbonic acid, the same stuff that is in carbonated beverages.

Activity 1 Key Terms

Materials Required
❑ Biology textbook or access to internet resources

Define the following terms:

Acidification

_____

_____

_____

Calcification

_____

_____

_____

Climate engineering

_____

_____

_____

Activity 2 Simulation of Ocean Acidification

Materials Required
❑ 1 head of red cabbage
❑ 2 quart saucepan
❑ 2 very small drinking cups (shot glasses or
   small juice glasses are best, plastic is fine)
❑ 2 drinking straws

1 Chop up the head of cabbage and boil it in the saucepan for 10 minutes (without a lid).

2 After it has cooled, pour a very small volume of the cabbage juice (about ½ inch or so) in the bottom of the cup, just enough that the end of the straw will be immersed.

3 Insert the straw all the way to the bottom of the cup and blow into the juice repeatedly for several minutes. The juice will turn purple when exposed to a base or pink when exposed to an acid.

4 Place an unused straw into the other cup, but do not blow into it. This will be your control.

5 What's the source of the CO_2 that you are exhaling through the straw?

6 Does the control cup change color? What color does the juice turn after you blow a bunch of CO_2 into it? Does it become more acidic or basic?

_____

_____

_____

Activity 3 Synthesis

Materials Required
No materials required for this activity

Now let's tie it all together.

1 What effect does ocean acidification have on marine organisms that have a calcified shell? How might changes in the abundance and distribution of shelled organisms affect marine ecosystems? (*Hint:* Think about trophic cascades and food webs.)

_____

_____

_____

_____

_____

_____

2 Pretend that you are a climate engineer hired by the United Nations to come up with a way of mitigating temperature and pH effects of carbon emissions on ocean acidification. Describe an approach that you would take to prevent or reverse ocean acidification.

_____

_____

_____

_____

_____

_____

Exercise 35
Ecosystem Ecology: Current Issues in Ecology

Objectives

At the completion of this exercise, students will be able to:

1 Define and apply key terms associated with current issues in ecology.

2 Discuss the pressures and factors affecting our environment.

3 Explain the acronym *HIPPO*.

4 Be aware of the ecological concerns in their areas.

5 Distinguish between descriptive and normative science.

6 Identify important ecosystem services and the economic importance of preserving biodiversity and ecosystem functioning.

7 Identify the tenets of ethical arguments for biodiversity conservation.

(i) Background Information

There are many ecological issues in the news today: Air quality, global climate change, land usage, and energy reserves all affect our lives. All living species—plant, animal, or other— have limits to what they can adapt. Many have evolved to such a high degree of specialization that even a small change will cause irreparable damage. Scientists have recently introduced the acronym *HIPPO* to describe five of the major issues affecting our world today.

H—Habitat loss. Habitat loss can occur due to many factors. Clearing habitat for commercial or economic reasons is a major cause. Fragmentation, degradation, and loss of habitat are the greatest threats to organisms today.

I—Invasive species. An invasive species may be introduced either deliberately or accidentally. In a best-case scenario,

they can become a nuisance. However, in a worst-case situation, they can produce a major ecological disaster. For an invasive species to have the ability to survive in a new environment, they generally have a strong ability to adapt. It is this ability that allows them to out-compete the native species. The zebra mussel invasion into the freshwaters of the U.S. is an example.

P—Pollution. Pollution of our water, air, and land is a major ecological issue and has been for many years. Pollution in any environment destroys organisms by fouling their ability to breath, metabolize, or find a suitable habitat in which to live. Pollution is the major contributing factor to climate change.

P—Population growth. Population growth has many implications. The most obvious is the physical use of habitat for housing, agriculture, or economic reasons. This development causes the displacement of local species that, in some situations, can't adapt quickly enough and are lost.

O—Overexploitation. The overexploitation of plants and animals by humans is an ongoing concern. An example of this is the Atlantic bluefin tuna. Some scientists estimate that the Atlantic population has decreased by more than 90% since 1960. It is also believed that the illegal wildlife and wildlife product trade is worth an annual $10 billion to poachers and smugglers.

Most of the issues we hear about today fall into one or more of these categories. Some of the categories overlap as well. For example, the overexploitation of our forests for lumber has resulted in large areas where forests have been clear-cut. This results in habitat loss for those organisms that were dependent on the forested areas.

Activity 1 Key Terms

Materials Required
❑ Biology textbook or access to internet resources

Define the following terms:

Invasive species

_____

_____

Endangered species

_____

_____

Threatened species

_____

_____

Keystone species

_____

_____

Exotic species

_____

_____

Habitat loss

_____

_____

Environment

_____

_____

Normative science

_____

_____

Descriptive science

_____

_____

Ecosystem services

_____

_____

Activity 2 **Mapping Environmental Impacts in Your Area**

Materials Required
❑ Access to internet resources
❑ Pencil or pen
❑ Notepad
❑ Ruler or straightedge

The purpose of this map is to document current ecological issues and how close or distant these issues are from where you live. This map can be quite basic.

1 Draw a map of your area in the space provided, and place your home in the center of the map.

2 Identify two areas in your town or city where habitat loss has taken place or is taking place. Mark the areas clearly on your map, and label the type of habitat affected.

3 Identify four invasive species, including at least one plant and animal. Mark the identified species on your map.

4 Identify five areas on your map where pollution is taking place, identifying at least three different types of pollution. Mark each area and label what type of pollution is taking place.

5 Identify two areas on your map where population growth is impacting the environment.

6 Locate an example of overexploitation on your map and label the type of overexploitation that is taking place.

Activity 3 Determining Descriptive versus Normative Science

Materials Required
❏ Access to internet resources

Just about all of the biology you have studied up to this point in your study of biology can be defined as descriptive science, standing in stark contrast with "normative science." You might think that normative science is mostly concerned with things like economics and politics, whereby practitioners establish what the economic goal is, and then use the tools of math and science to generate the political and economic policies that will achieve the goal. Policies can be judged as "good" or "bad" based on how well they achieve their objectives. But medicine and conservation biology are also appropriately described as normative science, because they too establish an ideal of what is "healthy" and then set that as a standard against which all other cases are judged. In both medicine and conservation biology, scientific observations are used to describe the situation (i.e., an individual or ecosystem's health) in order to inform a policy or plan that can restore the system to the "way it should be." This raises the question: Because ecosystems are dynamic, and because humans are a part of their ecosystems, what is a "normal" ecosystem? Is there such a thing as a "healthy" ecosystem?

1 Why do you think people might disagree on what a "healthy" ecosystem looks like?

_____

_____

_____

_____

2 What is the source of this conflict?

_____

_____

_____

_____

Activity 4 Investigating Biodiversity and Ecosystem Stability

Materials Required
☐ Access to internet resources

1 It has been argued that human health and happiness is completely dependent upon the preservation of biological diversity and functional ecosystems. In the space provided, make a list of ecosystem services that justify this claim. (*Note*: Consider ecosystem provisioning, including bioprospecting, water purification, flood control, sewage and pollution decomposition, cultural services, pollination, food production, etc.)

2 It has been argued that in addition to the economic and social benefits of biodiversity and ecosystem services, humans also have a purely ethical obligation to preserve the integrity of ecosystems. In the space provided, describe at least one argument that supports this position and supporting statements that validate it.

Activity 5 Synthesis

Materials Required
No materials required for this activity

Now let's tie it all together.

1 What is the difference between threatened and endangered species?

_____

_____

_____

_____

_____

_____

_____

2 Explain what the acronym *HIPPO* stands for.

_____

_____

_____

_____

_____

_____

_____

_____

_____

_____

_____

3 Explain two types of habitat loss impacting your area and what, if any, efforts are being made to resolve it.

_____

_____

_____

_____

_____

_____

_____

4 Why is there a concern about invasive species?

5 What effect have invasive species had in your area?

6 Explain three types of pollution in your area and what, if any, efforts are being made to resolve them.

7 What effect is population growth having in your area?

8 Describe overexploitation and discuss whether it is taking place in your area.

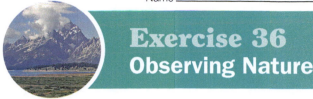

Exercise 36
Observing Nature

Objectives

At the completion of this exercise students will be able to:

1 Define and apply key terms associated with observing nature.

2 Record careful observations about their immediate surroundings in the natural world.

3 Ascribe their environment to an ecosystem type or biome.

4 Describe the climate parameters of their ecosystem and the adaptive response of organisms to these environmental drivers.

5 Diagram a food web and trophic pyramid based on observations of organisms in their natural environment.

6 Devise testable hypotheses for how the organisms in their observational study interact.

7 Generate testable hypotheses for exploring the relationship between time spent in natural environments and individual human well-being.

ⓘ Background Information

One of the reasons your biology instructors ended up teaching this course undoubtedly comes from the fact that somewhere along the line, they had an experience in nature that instilled a sense of wonder and awe that led them on a path of searching out a better understanding of the natural world. In this exercise, you will spend time in a natural setting and make observations about the natural world. When considering where to undertake the activities in this exercise, please try to find a place where you feel safe, but which is as undisturbed and wild as possible. Be prepared for weather (rain, wind, heat, cold, etc.)—don't back out of the assignment just because it isn't a perfect day! Turn off all sources of distraction (cell phones, MP3 devices, etc.) for the duration of each activity.

Activity 1 Key Terms

Materials Required
☐ Biology textbook or access to internet resources

Define the following terms:

Biophilia hypothesis

_____

_____

Consilience

_____

_____

Nature deficit disorder

_____

_____

Deep ecology

_____

_____

Environmental stewardship

_____

_____

Environmental ethics

_____

_____

_____

Terrestrial ecosystem types:

Taiga

_____

_____

Tundra

_____

_____

Deciduous forest

_____

_____

Grassland

_____

_____

Tropical rain forest

_____

_____

Desert

_____

_____

Name _____ Date _____ Section _____

Activity 2 Making Observations in Nature

Materials Required
- ❑ Pencil or pen (for recording observations)
- ❑ Sound recording device (optional)
- ❑ Colored pencils, markers, watercolors, paint, etc. (optional)

In this activity, you will spend one hour alone in a natural, outdoor setting recording what you can sense (see, feel, hear, and touch). Before embarking on this activity, be sure to complete Activity 1 and familiarize yourself with the key terms. Make as little noise as possible, but feel free to move about and observe your surroundings. This could include looking up, looking down, turning over leaves and rocks, digging up handfuls of soil with your hands, etc.

1 In the space provided, sketch five different plant species that you encounter in your area. Be sure to include details such as their leaf structure, surface texture, color, branching patterns, stems, leaves, needles, flowers, and seeds.

2 In the space below, sketch three different animal species that you see.

3 Make a sound recording, or list and describe the noises you hear. Identify an organism that may be responsible for each sound.

Activity 3 Placing Organisms in an Ecological Context

Materials Required
- ☐ Pencil or pen (for recording observations)
- ☐ Colored pencils, markers, watercolors, paint, etc. (optional)

Biologists can learn a lot by observing organisms in a Petri dish, on a microscope slide, or mounted on an insect pin. However, these organisms reveal so much more about their biology when observed in their natural environment. This activity will require you to go outdoors and explore organisms and their ecosystem.

1 Using your definitions from Activity 1, identify what type of terrestrial ecosystem you are observing.

_____

_____

2 Describe the following aspects of your ecosystem. Include in your description the current conditions (i.e., current temperature) as well as the mean annual condition, or climate.

   a. Temperature

_____

_____

_____

   b. Precipitation

_____

_____

_____

   c. Sunlight

_____

_____

_____

   d. Wind

_____

_____

_____

3 Describe, in general terms, the four components of the ecosystem you are observing.

   a. Abiotic environment

_____

_____

_____

b. Producers

_____

_____

_____

c. Consumers

_____

_____

_____

d. Decomposers

_____

_____

_____

4 Describe how the climate drivers in Questions 2 and 3 have shaped at least one aspect of each of the plants and animals you sketched in Activity 2. Consider the role of adaptation in your answer.

a. Plant 1

_____

_____

_____

b. Plant 2

_____

_____

_____

c. Plant 3

_____

_____

_____

d. Plant 4

_____

_____

_____

_____

e. Plant 5

_____

_____

_____

_____

f. Animal 1

_____

_____

_____

_____

g. Animal 2

_____

_____

_____

_____

h. Animal 3

_____

_____

_____

_____

5 Draw a food web that includes the plants and animals you described in step 4.

6 Diagram a trophic pyramid that indicates the trophic levels in the ecosystem you are observing. On your diagram, indicate where each of the organisms exist.

7 Identify the functional role of each organism you sketched. Consider roles such as primary producer, primary consumer, decomposer, secondary consumer, tertiary consumer, etc.

8 List the biotic interactions that may exist among the organisms you sketched. Do you have any predator-prey relationships? Host-parasite? Commensal? Host-pathogen? Competition?

9 Choose one of the biotic interactions in Question 8, and sketch a graph of how they might influence each other's growth curves.

Activity 4 Your Place in Nature

Materials Required
❑ Pencil or pen (for recording observations)

To a biologist, it is unfortunate that human beings, especially children and young adults, are spending increasingly more time indoors away from the natural world. This trend has been linked to a wide range of social and behavioral disorders. In recent years, the term *nature deficit disorder* has been used to describe the correlation between social and behavioral disorders and time spent in nature. Thus, one of the objectives of this assignment is to ponder the idea of how spending time in nature is linked to your own perspectives on mental health.

1 While observing nature, take 15 minutes to simply observe your surroundings in silence. Don't take notes or measurements. Clear your head of things like assignment due dates, relationships with others, or other sources of stress and distraction. Just "be" in your environment. Using appropriate terms from this exercise, write down a few thoughts or feelings you have about your experience.

2 In addition to learning more about the world around us by spending time in nature, it has also been hypothesized that spending time in natural environments can also alleviate problems such as obesity, ADD, and depression. Spending time in the wild has also been argued as a way to improve an individual's creativity, cognitive development, and emotional well-being. Choosing one of these aspects, develop a testable hypothesis and explanation for how this could be. Be sure to include appropriate terms from this lesson in the design of your experiment.

Activity 5 Synthesis

Materials Required
No materials required for this activity

Now let's tie it all together.

1 Using the appropriate terms from this lesson, describe one new biological insight that you gained from your experience observing nature.

2 Did your experience provide any insight into how spending time outdoors contributes to your overall well-being? Explain your answer using the appropriate terms from this exercise.

Index

redox reaction, 89 (Fig. 10.1), 92
respiration
    aerobic, 263
    comparison with photosynthesis, 105
      (Fig. 11.5)
    in goldfish, 263
    physiological, 263
Rh factor/inheritance, 137 (Table 15.1)
RNA (mRNA, rRNA, and tRNA), 155, 159
    (Fig. 17.4), 161
    characteristics, 183
    comparison with DNA, 157 (Fig. 17.2)
    hypotheses, 183
    polymerase, 161, 174
    polynucleotides, 183
    and ribosome, 158 (Fig. 17.3)
rubisco, 102

science
    descriptive, 310
    difference from pseudoscience, 7
    normative, 310

scientific experiments, 19
scientific method, 19, 27
scientific principles, formulation of a
    question, 19
selectively permeable membrane, 67
sex-linked traits, 133
skepticism versus denialism, 7
species concept/speciation, 217
    initial allele frequencies, 221 (Table 25.1)
    "North" population, 222 (Table 25.2)
    "South" population, 222 (Table 25.3)
square shape, four straws connected, 71
    (Fig. 8.3)
sugar, 95
    making, 101
sulfur, 37

teeth, wisdom, 196 (Fig. 22.3)
testing, as scientific principle, 19
Tetrapoda, 206 (Fig. 23.2)
three-domain system, 231, 232 (Table 26.1)
thylakoid membrane, 101

Tinbergen, Niko, 358
transcription, 161, 163 (Fig. 18.1)
translation, 161, 164 (Fig. 18.2)
tree (photo), 5
tree of life, 232, 233 (Fig. 26.1)
triglyceride, 52 (Fig. 6.6)
tRNA structure, 157 (Fig. 17.2), 158
    (Fig. 17.3)
turgor pressure, 77

water
    Earth, saltwater and freshwater, 291
    properties, 43, 55 (photo), 56 (photo)
Watson, James, DNA model, 147
Western Bluebird (*Sialia mexicana*)
    (Figs. 24.1, 24.2), 213
Wilkins, Maurice, 147
Woese, Carl (three domain system), 231
world map, 197 (Fig. 22.4)

yeast, alcohol formation, 89